ENCYCLOPAEDIA OF
PLANT PHYSIOLOGY

ENCYCLOPAEDIA OF PLANT PHYSIOLOGY

Dr. Satyavir Singh

ANMOL PUBLICATIONS PVT. LTD.
NEW DELHI - 110 002 (INDIA)

ANMOL PUBLICATIONS PVT. LTD.
H.O.: 4374/4B, Ansari Road, Darya Ganj,
New Delhi-110 002 (India)
Ph.: 23278000, 23261597

B.O.: No. 1015, Ist Main Road, BSK IIIrd Stage
IIIrd Phase, IIIrd Block
Bangalore - 560 085 (India)
Visit us at: www.anmolpublications.com

Encyclopaedia of Plant Physiology

First Published, 2006

ISBN 81-261-3033-4

PRINTED IN INDIA

Printed at Mehra Offset Press, Delhi.

Preface

This Encyclopaedia is the perfect guide for those studying Plant physiology, either at university level or anyone interested in Plant physiology, such as plant pathologists. It provides a comprehensive coverage of Plant physiology. The aim of this encyclopaedia is to provide teachers, students and laymen interested in Plant physiology with clear, concise, and correct definitions and descriptions of the Plant physiology terms, throughout.

In this volume all the branches of Plant physiology are represented. This volume covers whole realm of Plant physiology and includes various topics such as activities in cells, tissues, and organs, processes such as contractility of muscle tissue, coordination through the nervous system, feeding, digestion, excretion, respiration, circulation, reproduction, and secretion.

The Encyclopaedia is to satisfy the immediate technical needs of the user. Concepts are presented in a clear and lucid manner. In terms of its broader objectives, the present book may be viewed as an outline or compendium of Plant physiology ideas correlated with a large collection of specialized items.

The aim of this volume is to provide a concise repository of number of major concepts in this field. In preparation of this book, the author has freely consulted large number of books and journals so no authenticity is claimed. Author is especially thankful to Anmol Publications Pvt. Ltd., New Delhi for shaping this book in its final form. Suggestions for further improvement of this book are not only welcome but also greatly appreciated.

Author

A

A horizon Mineral horizons which formed at the surface or below an O horizon and which exhibit obliteration of all, or much, of the original rock structure and are: 1. Characterized by an accumulation of humified organic matter intimately mixed with the mineral fraction and not dominated by properties characteristic of E or B horizons; or

2. Have properties resulting from cultivation, pasturing, or similar kinds of disturbance.

ABA (Abscisic Acid) A plant growth regulator found in plants that acts as a signaling molecule which helps to control overall growth and physiology. It regulates responses to environmental stresses such as drought, cold, salinity and heat.

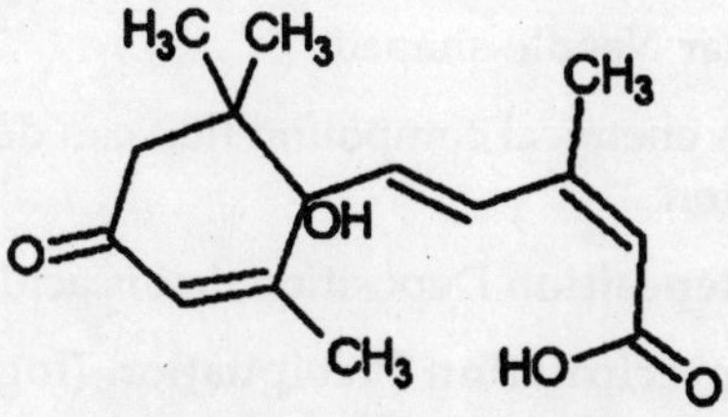

Fig. ABA (Abscisic Acid)

Abaxial Directed away from the stem of a plant; pertaining to the lower surface of a leaf.

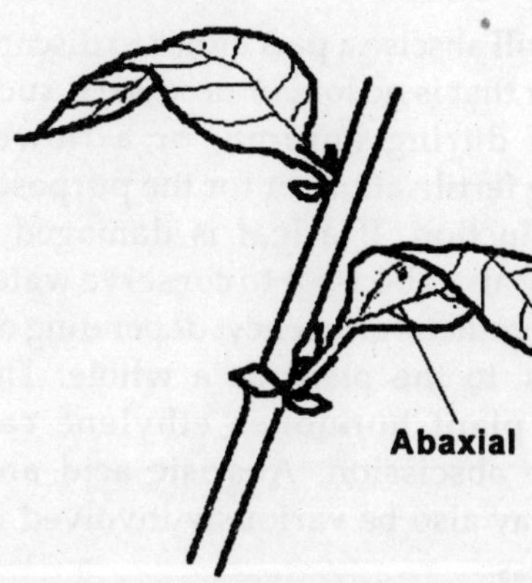

Fig. Abaxial

ABC soil Soil that has a surface layer, subsoil, and a lower layer of loose rock material is called this. The layers are said to be it's A, B, and C horizons.

ABC transporters A large family of membrane transport proteins characterized by a highly conserved ATP binding domain.

Aberrant This describes a plant or species that is different in some way from the group in which it is placed.

Abiotic Pertaining to physical and inorganic components. For example, diseases/ disorders in plants can be caused by abiotic factors such as extremes of heat, light, moisture, lack of nutrients etc.

Ablastous Means without germ or bud.

Abruptly pinnate This describes a pinnate leaf that ends without an odd leaflet or tendril. It ends in a matched pair of leaflets.

Abscise To separate from a plant, as leaves, flowers, and fruits do when they fall.

Abscission Abscission (from ab- away from, and scission cutting or severing) is the shedding of a body part. It most commonly refers to the process by which a plant intentionally drops one or more of its parts, such as a leaf, fruit, flower or seed, though the term is also used to describe the shedding of a claw by an animal.

A plant will abscise a part either to discard a member that is no longer necessary, such as a leaf during autumn, or a flower following fertilisation, or for the purposes of reproduction. If a leaf is damaged a plant may also abscise it to conserve water or photosynthetic efficiency, depending on the 'costs' to the plant as a whole. The gaseous plant hormones ethylene can stimulate abscission. Abscisic acid and auxins may also be variously involved in the process.

Absorption The process of moving a substance from the exterior of a plant to the internal portion.

Acantha A botanical word that refers to a spine, thorn or prickle.

Acarpous This describes a plant that is sterile.

Acaulescent A plant that is stemless. The word Acauline is also used.

ACC oxidase The enzyme in the ethylene biosynthesis pathway converting ACC (1-aminocyclopropane-1-carboxylic acid) to ethylene.

ACC synthase The enzyme in the ethylene biosynthesis pathway converting SAM (S-adenosyl-methionine) to ACC (1-aminocyclopropane-1-carboxylic acid).

ACCase Acetyl-Coenzyme A carboxylase; enzyme responsible for production of fatty acids and is the target of many herbicides that have the lipid synthesis inhibitor mode of action.

Acceptable daily intake (ADI) A measure of the level of a chemical, i.e. a pesticide residue, that is believed to be able to be consumed on a daily basis over a lifetime without harm; used by the U.S. government to establish safe levels of potentially hazardous substances in food.

Accretive Increasing or adding by growth; continued development within.

Accumbent Lying against. Mainly used to describe the first sprouts of an embryo when they lie against the body of the seed.

Aceae New Latin suffix, meaning "the plants of the nature of", which is used to form names of families of plants. Examples are Liliaceae, Rosaceae, etc.

Acephate A man-made, systemic insecticide that is used to control numerous pests such as aphids, scale and thrips, of most ornamental plants.

Acervate Growing in heaps or in compact clusters.

Acervulus Erumpent, cushionlike fruiting body bearing conidiophores, conidia, and sometimes setae.

Achene This is a small, dry, hard, one-celled, one-seeded fruit that stays closed at maturity.

Acicular Needle-shaped.

Acid A chemical compound that can donate a proton

Acid deposition Deposition that is acidic.

Acid precipitation Precipitation (fog, rain, snow) with a low pH, due to the presence of nitric and sulfuric acid formed by the reaction of air pollutants ($N0x$ and $S0_2$) with water.

Acid rain Acid rain is normally considered to be a by-product of modern atmospheric pollution. Even in a pure, uncontaminated world, however, it is likely that the rainfall would be acidic. Carbonic acid (H_2CO_3) is formed when carbon dioxide (CO_2) is absorbed by atmospheric water. Nitric acid (HNO_3) is created during thunderstorms and sulphuric acid (H_2SO_4) is formed from the sulphur dioxide (SO_2) released during volcanic eruptions and forest fires or from the sulphur (S) emitted by phytoplankton during their seasonal bloom period. All of these are natural processes which contribute to make normal rain acid. Current concern over acid rain is not with the naturally produced variety, but rather with that which results from modern industrial activity and the acid gases it produces. Considerable amounts of sulphur dioxide are released into the atmosphere as a by-product of metal smelting, particularly when non-ferrous ores are involved. The burning of coal and oil to provide energy for space heating or to fuel thermal electric power stations also produces sulphur dioxide. The continuing growth of transportation systems using the internal combustion engine contributes to acid rain through the release of oxides of nitrogen (NO_x) into the atmosphere.

Acid precipitation produced by human activities differs from natural acid precipitation not only in its origins, but also in its quality. Anthropogenically produced acid rain tends to be many times more acidic than the natural variety. The difference is commonly of the order of 1.0 to 1.5 points on the pH scale used to measure acidity. In North America, for example, naturally acid rain has a pH of about 5.6, whereas measurements of rain falling in southern Ontario, Canada frequently provide values in the range of 4.5 to 4.0. To put these values in perspective, vinegar has a pH of 2.7 and milk a pH of 6.6. Similar values for background levels of acidic rain are indicated by studies in Europe with the average annual pH of rain over Britain between 1978 and 1980 being within the range of 4.5 to 4.2. Acid precipitation produced by human activities differs from natural acid precipitation not only in its origins, but also in its quality. Anthropogenically produced acid rain tends to be many times more acidic than the natural variety. The difference is commonly of the order of 1.0 to 1.5 points on the pH scale used to measure acidity. In North America, for example, naturally acid rain has a pH of about 5.6, whereas measurements of rain falling in southern Ontario, Canada frequently provide values in the range of 4.5 to 4.0. To put these values in perspective, vinegar has a pH of 2.7 and milk a pH of 6.6. Similar values for background levels of acidic rain are indicated by studies in Europe with the average annual pH of rain over Britain between 1978 and 1980 being within the range of 4.5 to 4.2.

The quality of the rain is determined by a series of chemical processes set in motion when acidic materials are released into the atmosphere. Some of the sulphur dioxide and oxides of nitrogen emitted will return to the surface quite quickly, and close to their source, as dry deposition. The remainder will be carried up into the atmosphere, to be converted into sulphuric and nitric acid, which will eventually return to earth as acid rain. The processes involved are fundamentally simple. Oxidation converts the gases into acids, in either a gas or liquid phase reaction, at a rate which depends upon such variables as the concentration of heavy metals in airborne particulate matter, the presence of ammonia and the intensity of sunlight. The metals and ammonia appear to act as catalysts, while the sunlight provides the energy required by the chemical reactions. The quality of the rain is determined by a series of chemical processes set in motion when acidic materials are released into the atmosphere. Some of the sulphur dioxide and oxides of nitrogen emitted will return to the surface quite quickly, and close to their source, as dry deposition. The remainder will be carried up into the atmosphere, to be converted into sulphuric and nitric acid, which will eventually return to earth as acid rain. The processes involved are fundamentally simple. Oxidation converts the gases into acids, in either a gas or liquid phase reaction, at a rate which depends upon such variables as the concentration of heavy metals in airborne particulate matter, the presence of ammonia and the intensity of sunlight. The metals and ammonia appear to act as catalysts, while the sunlight provides the energy required by the chemical reactions.

Acid rain remained mainly a local problem in the past. The introduction of the tall stacks policy in an attempt to reduce local pollution, however, increased its geographical extent. The release of pollutants at the greater altitudes possible with the taller smokestacks placed them outside the boundary layer circulation and into the larger scale atmospheric circulation system with its potential for much greater dispersal through the mechanisms involved in the Long Range Transportation of Air Pollution (LRTAP). Local problems were mitigated at the expense of the larger environment. Acid rain remained mainly a local problem in the past. The introduction of the tall stacks policy in an attempt to reduce local pollution, however, increased its geographical extent. The release of pollutants at the greater altitudes possible with the taller smokestacks placed them outside the boundary layer circulation and into the larger scale atmospheric circulation system with its potential for much greater dispersal through the mechanisms involved in the Long Range Transportation of Air Pollution (LRTAP). Local problems were mitigated at the expense of the larger environment.

The main sources of acid rain are to be found in the industrialized areas of the northern hemisphere. North-eastern North America, Britain and Western Europe have received most attention, although their output of sulphur dioxide has been declining. Levels of oxides of nitrogen have not yet experienced significant decline and continue to rise in some areas. The emission of acid gases from Eastern Europe and the republics of the former USSR - Russia, Ukraine and Kazakhstan - remains high. In Asia, Japanese industries emit large quantities of sulphur dioxide, while the industrial areas of China are also major contributors. Acid emissions remain limited outside of the major industrial nations, but concern has been expressed over growing levels of air pollution, often associated with urban automobile exhaust emissions, which may already have provided a base for acid rain in some Third World countries. The main sources of acid rain are to be found in the industrialized areas of the northern hemisphere. North-eastern North America, Britain and Western Europe have received most attention, although their output of sulphur dioxide has been declining. Levels of oxides of nitrogen have not yet experienced significant decline and continue to rise in some areas. The emission of acid gases from Eastern Europe and the republics of the former USSR - Russia, Ukraine and Kazakhstan - remains high. In Asia, Japanese industries emit large quantities of sulphur dioxide, while the industrial areas of China are also major contributors. Acid emissions remain limited outside of the major industrial nations, but concern has been expressed over growing levels of air pollution, often associated with urban automobile exhaust emissions, which may already have provided a base for acid rain in some Third World countries.

Once the acid gases have been released into the atmosphere, they are at the mercy of the prevailing circulation patterns. With almost all of the areas currently producing large amounts of acidic pollution located within the mid-latitude westerly wind belt, emissions are normally carried eastwards, or perhaps north-eastwards, often for several hundred kilometres before being redeposited. In this way, pollutants originating in the US Midwest cause acid rain in Ontario, Quebec and the New England states. Emissions from the smelters and power stations of the English Midlands and the Ruhr contribute to the acidity of precipitation in Scandinavia, and acidity in the Arctic originates as far as 8000 km away to the south in North America and Eurasia. Thus the problem of acid rain transcends national boundaries, introducing political overtones to the problem and creating the need for international cooperation if a solution is to be found. Once the acid gases have been released into the atmosphere, they are at the mercy of the prevailing circulation patterns. With almost all of the areas currently producing large amounts of acidic pollution located within the mid-latitude westerly wind belt, emissions are normally carried eastwards, or perhaps north-eastwards, often for several hundred kilometres before being redeposited. In this way, pollutants originating in the US Midwest cause acid rain in Ontario, Quebec and the New England states. Emissions from the smelters and power stations of the English Midlands and the Ruhr contribute to the acidity of precipitation in Scandinavia, and acidity in the Arctic originates as far as 8000 km away to the south in North America and Eurasia. Thus the problem of acid rain transcends national boundaries, introducing political overtones to the problem and creating the need for international cooperation if a solution is to be found.

The impact of acid rain on the environment depends not only on the level of acidity in the rain, but also on the nature of the environment itself. Areas underlain by acidic rocks such as granitic or quartzitic bedrock, for example, are particularly susceptible to damage, lacking as they do the ability to 'buffer' or neutralize additional acidity from the precipitation. Acid levels therefore rise, the environmental balance is disturbed and serious ecological damage is the inevitable result. In contrast, areas which are geologically basic - underlain by limestone or chalk, for example - are much less sensitive, and may even benefit from the additional acidity. The highly alkaline soils and water of areas underlain by chalk or limestone ensure that the acid added to the environment by the rain is very effectively neutralized. The impact of acid rain on the environment depends not only on the level of acidity in the rain, but also on the nature of the environment itself. Areas underlain by acidic rocks such as granitic or quartzitic bedrock, for example, are particularly susceptible to damage, lacking as they do the ability to 'buffer' or neutralize additional acidity from the precipitation. Acid levels therefore rise, the environmental balance is disturbed and serious ecological damage is the inevitable result. In contrast, areas which are geologically basic - underlain by limestone or chalk, for example - are much less sensitive, and may even benefit from the additional acidity. The highly alkaline soils and water of areas underlain by chalk or limestone ensure that the acid added to the environment by the rain is very effectively neutralized.

The impact of acid rain on the environment was first recognized in the lakes and rivers of these areas. Reduced pH values which indicated the rising acidity were accompanied by low levels of calcium (Ca) and magnesium (Mg), elevated sulphate concentrations and an increase in the amounts of potentially toxic metals such as aluminum (Al). When aquatic communities were examined in areas as far apart as New York State, Nova Scotia, Norway and Sweden there was clear evidence that increased surface water acidity had adverse effects on fish. In some cases the acidity was sufficiently high that mature fish died, but, more commonly, fish populations began to decline because of the effects of the increasing acidity on reproduction. Damage to the eggs during spawning and the inability of the young fry to survive the higher acidity, particularly during the spring flush, ensured that the older fish were not replaced as they died. As a result, fish populations in many rivers and lakes in eastern North America, Britain and Scandinavia have declined noticeably in the last two to three decades and hundreds of lakes are now completely devoid of fish. Although the impact on the fish is most obvious, all aquatic organisms will decline in number and variety during progressive acidification. Waterbodies that have lost or are in the process of losing their flora and fauna are often described as 'dead' or 'dying'. This is not strictly correct, however, for some organisms are remarkably acid-tolerant, and even the most acid lakes have some life in them. Species of protozoans are found at pH levels as low as 2.0, for example.

The impact of acid rain on the environment was first recognized in the lakes and rivers of these areas. Reduced pH values which indicated the rising acidity were accompanied by low levels of calcium (Ca) and magnesium (Mg), elevated sulphate concentrations and an increase in the amounts of potentially toxic metals such as aluminum (Al). When aquatic communities were examined in areas as far apart as New York State, Nova Scotia, Norway and Sweden there was clear evidence that increased surface water acidity had adverse effects on fish. In some cases the acidity was sufficiently high that mature fish died, but, more commonly, fish populations began to decline because of the effects of the increasing acidity on reproduction. Damage to the eggs during spawning and the inability of the young fry to survive the higher acidity, particularly during the spring flush, ensured that the older fish were not replaced as they died. As a result, fish populations in many rivers and lakes in eastern North America, Britain and Scandinavia have declined noticeably in the last two to three decades and hundreds of lakes are now completely devoid of fish. Although the impact on the fish is most obvious, all aquatic organisms will decline in number and variety during progressive acidification. Waterbodies that have lost or are in the process of losing their flora and fauna are often described as 'dead' or 'dying'. This is not strictly correct, however, for some organisms are remarkably acid-tolerant, and even the most acid lakes have some life in them. Species of protozoans are found at pH levels as low as 2.0, for example.

There is growing evidence that those areas in which the waterbodies have already succumbed to acidification must also face the effects of increasing acid stress on their forests and soils. The threat is not universally recognized, however, and there remains a great deal of controversy over the amount of damage directly attributable to acid rain. Reduction in forest growth in Sweden, physical damage to trees in West Germany, and the death of sugar maples in Quebec and Vermont have all been blamed on the increased acidity of the precipitation in these areas. Many of the impacts, such as the thinning of annual growth rings, reduction in biomass and damage to fine root systems, are only apparent after detailed examination, but others are more directly obvious and have been described as dieback. This involves the gradual wasting of the tree inwards from the outermost tips of its branches. The process is cumulative over several years until the tree dies. The symptoms of dieback have been recognized in the maple groves and red spruce forests of north-eastern North America, and across fifteen countries and some 70,000 km^2 of forest in Europe. In Germany, where it was first linked with acid rain, dieback, or Waldsterben, was particularly extensive and in the 1980s was seen as a major threat to the survival of the German forests. More recent research has indicated that forest decline in Europe is a multifaceted process in which acid precipitation may be only one of a series of contributors along with tree harvesting practices, drought and fungal attacks, although its exact role remains controversial. There is growing evidence that those areas in which the waterbodies have already succumbed to acidification must also face the effects of increasing acid stress on their forests and soils. The threat is not universally recognized, however, and there remains a great deal of controversy over the amount of damage directly attributable to acid rain. Reduction in forest growth in Sweden, physical damage to trees in West Germany, and the death of sugar maples in Quebec and Vermont have all been blamed on the increased acidity of the precipitation in these areas. Many of the impacts, such as the thinning of annual growth rings, reduction in biomass and damage to fine root systems, are only apparent after detailed examination, but others are more directly obvious and have been described as dieback. This involves the gradual wasting of the tree inwards from the outermost tips of its branches. The process is cumulative over several years until the tree dies. The symptoms of dieback have been recognized in the maple groves and red spruce forests of north-eastern North America, and across fifteen countries and some 70,000 km2 of forest in Europe. In Germany, where it was first linked with acid rain, dieback, or Waldsterben, was particularly extensive and in the 1980s was seen as a major threat to the survival of the German forests. More recent research has indicated that forest decline in Europe is a multifaceted process in which acid precipitation may be only one of a series of contributors along with tree harvesting practices, drought and fungal attacks, although its exact role remains controversial.

Present concern over acid rain is concentrated mainly on its effect on the natural environment, but acid rain also contributes to deterioration in the built environment: it attacks limestone and marble used as building stone, for example. The faceless statues and crumbling cornices of the world's famous palaces, castles, abbeys and cathedrals, from the Parthenon in Greece to the Taj Mahal in India, attest to its power. By attacking the fabric of these buildings it not only causes physical and economic damage, but threatens the world's cultural heritage. Present concern over acid rain is concentrated mainly on its effect on the natural environment, but acid rain also contributes to deterioration in the built environment: it attacks limestone and marble used as building stone, for example. The faceless statues and crumbling cornices of the world's famous palaces, castles, abbeys and cathedrals, from the Parthenon in Greece to the Taj Mahal in India, attest to its power. By attacking the fabric of these buildings it not only causes physical and economic damage, but threatens the world's cultural heritage.

Acid rain in the form of aerosols or attached to smoke particles can cause respiratory problems in humans, as was the case with the infamous London Smog of 1952. There is no evidence that wet deposition is directly damaging to human health, although in its ability to mobilize metals, from lead (Pb) or copper (Cu) pipes, for example, it may have important indirect effects. Acid rain in the form of aerosols or attached to smoke particles can cause respiratory problems in humans, as was the case with the infamous London Smog of 1952. There is no evidence that wet deposition is directly damaging to human health, although in its ability to mobilize metals, from lead (Pb) or copper (Cu) pipes, for example, it may have important indirect effects.

Solutions to the problem of acid rain are deceptively simple. In theory, a reduction in the emission rate of acid forming gases is all that is required to slow down and eventually stop the damage being caused by the acidification of the environment. Translating that concept into reality has proved difficult, however. One of the first approaches involved adaptation, such as the addition of lime to land or lakes to counteract the increased acidity, and to allow the recovery mechanisms to work more effectively. Experiments involving the liming of lakes in Canada and Sweden, for example, have provided encouraging results. Artificial buffering of lakes in this way is only a temporary measure which may be likened to the use of antacid to reduce acid indigestion. The neutralizing effects of the lime may last longer than those of the antacid, but they do wear off in three to five years and re-liming is necessary as long as acid loading continues. Most of the current proposals for dealing with acid rain tackle the problem at its source. They attempt to prevent, or at least reduce, emissions of acid gases into the atmosphere, and since sulphur dioxide makes the greatest contribution to acid rain in North America and Europe, it has received most attention in the development of abatement procedures. Sulphur dioxide is formed when coal and oil are burned to release energy, and the technology to control it may be applied before, during or after combustion. The exact timing will depend upon such factors as the amount of acid reduction required, the type and age of the system and the cost-effectiveness of the particular process.

Solutions to the problem of acid rain are deceptively simple. In theory, a reduction in the emission rate of acid forming gases is all that is required to slow down and eventually stop the damage being caused by the acidification of the environment. Translating that concept into reality has proved difficult, however. One of the first approaches involved adaptation, such as the addition of lime to land or lakes to counteract the increased acidity, and to allow the recovery mechanisms to work more effectively. Experiments involving the liming of lakes in Canada and Sweden, for example, have provided encouraging results. Artificial buffering of lakes in this way is only a temporary measure which may be likened to the use of antacid to reduce acid indigestion. The neutralizing effects of the lime may last longer than those of the antacid, but they do wear off in three to five years and re-liming is necessary as long as acid loading continues. Most of the current proposals for dealing with acid rain tackle the problem at its source. They attempt to prevent, or at least reduce, emissions of acid gases into the atmosphere, and since sulphur dioxide makes the greatest contribution to acid rain in North America and Europe, it has received most attention in the development of abatement procedures. Sulphur dioxide is formed when coal and oil are burned to release energy, and the technology to control it may be applied before, during or after combustion. The exact timing will depend upon such factors as the amount of acid reduction required, the type and age of the system and the cost-effectiveness of the particular process.

Flue gas desulphurization (FGD) is the most common approach to sulphur dioxide reduction. It uses scrubbers to neutralize the acidity of the flue gases before they are released into the atmosphere. With sulphur dioxide removal rates of between 80 and 95 per cent, FGD is more effective than methods such as fuel switching or fuel desulphurization applied prior to combustion and it is technically simpler than fluidized bed combustion (FBC) and lime injection multi-stage burning (LIMB) incorporated in the combustion process itself. Although FGD systems do little to reduce emissions of oxides of nitrogen they are the preferred method of dealing with acid rain, and future requirements to reduce sulphur dioxide levels will be met in large part by the installation of FGD equipment. Flue gas desulphurization (FGD) is the most common approach to sulphur dioxide reduction. It uses scrubbers to neutralize the acidity of the flue gases before they are released into the atmosphere. With sulphur dioxide removal rates of between 80 and 95 per cent, FGD is more effective than methods such as fuel switching or fuel desulphurization applied prior to combustion and it is technically simpler than fluidized bed combustion (FBC) and lime injection multi-stage burning (LIMB) incorporated in the combustion process itself. Although FGD systems do little to reduce emissions of oxides of nitrogen they are the preferred method of dealing with acid rain, and future requirements to reduce sulphur dioxide levels will be met in large part by the installation of FGD equipment.

The study of acid rain has declined remarkably since the 1980s, when it was viewed by many as the major environmental problem facing the northern hemisphere. It is one environmental issue in which abatement programmes have met with some success. Sulphur dioxide levels continue to decline, the rain in many areas is considerably less acid than it was a decade ago, and the transboundary disputes which absorbed large amounts of time, energy and money in the 1980s have been resolved. There are indications that lakes and forests are showing some signs of recovery in the most vulnerable areas of North America and Europe, although this is still a matter of dispute. The study of acid rain has declined remarkably since the 1980s, when it was viewed by many as the major environmental problem facing the northern hemisphere. It is one environmental issue in which abatement programmes have met with some success. Sulphur dioxide levels continue to decline, the rain in many areas is considerably less acid than it was a decade ago, and the transboundary disputes which absorbed large amounts of time, energy and money in the 1980s have been resolved. There are indications that lakes and forests are showing some signs of recovery in the most vulnerable areas of North America and Europe, although this is still a matter of dispute.

Developments such as these have created the perception that the acid rain problem is being solved. It would be more accurate, however, to say that it has changed in nature and in geographical extent. In the developed nations, oxides of nitrogen are gradually making a larger contribution to acidity as levels of sulphur dioxide decline. In less developed areas, from Eastern Europe to China, the full impact of acid rain may still be in the future, and the research activities currently winding down in Western Europe and North America may have to be revived to deal with it.

Acid soil Soil with a pH value of less than 7, or lacking in lime.

Acidic Having a pH lower than 7, having a higher concentration of H+ ions.

Acidification The process of becoming acidic. For example, as acid rain falls on a lake over the many years, the pH of the lake itself may drop (become more acidic); this process is called acidification.

Acidify To lower the soil's pH to meet the level of acid soil.

Acidity The level of acid prescent in the soil.

Acid-neutralizing capacity (ANC) A measure of the ability of water or soil to neutralize added acids, such as from acid deposition. Streams with higher ANC levels are affected less by acid rain than streams with lower ANC values.

Acinaceous Consisting of or full of kernels.

Acinacifolious Having scimitar-shaped leaves.

Acinaciform Shaped like a scimitar (usually describes a leaf shape). An acinaciform leaf has one edge convex and thin and the other edge straighter and thick. This term is also used to describe the shape of the pods of some beans.

Acinarious Covered with little round, stalked, grape-like sacs, as in some algae.

Aciniform Having the form of grapes, or growing in grape-like clusters; acinose.

Acrogen A plant that increases by growth at the summit or terminal buds only. The acrogens form a division of the Cryptogams, including ferns, fern allies, liverworts, and mosses.

Acropetal Upward from the base to the apex of a shoot of a plant; in fungi, the production of spores in succession in the direction of the apex so that the apical spore is the youngest.

Acrylamide Chemical with the basic formula C_3H_5NO that can be chemically induced to form a gel matrix. This matix can be used for electrophoretic separation of proteins and nucleic acids.

Actin Actin is a globular structural protein that polymerizes in a helical fashion to form an actin filament (or microfilament). These form the cytoskeleton - a three-dimensional network inside an eukaryotic cell. Actin filaments provide mechanical support for the cell, determine the cell shape, enable cell movements (through lamellipodia, filopodia, or pseudopodia); and participate in certain cell junctions, in cytoplasmic streaming and in contraction of the cell during cytokinesis. In muscle cells they play an essential role, along with myosin, in muscle contraction. In the cytosol, actin is predominantly bound to ATP, but can also bind to ADP. An ATP-actin complex polymerizes faster and dissociates slower than an ADP-actin complex. Actin is one of the most abundant proteins in many eukaryotic cells, with concentrations of over 100 μM. It is also one of the most highly conserved proteins, differing by no more than 5% in species as diverse as algae and humans.

Actin bundle Actin filaments that are crosslinked into closely packed arrays.

Actin network Actin filaments that are crosslinked into loose three-dimensional meshworks.

Actinomycete A group of microorganisms similar to bacteria that produce long filaments.

Action site A location in dense vegetation, that will collect water droplets and ensure a high degree of moth survival. The area where mating between male and female moths occurs.

Activation energy The energy required to raise a molecule to its transition state to undergo a chemical reaction.

Activator Any substances, such as microorganisms and nutrients, which are added to a compost pile to speed the breakdown of organic matter. Also called compost activator and compost inoculant.

Active site The location on an enzyme (protein) to which substrate molecules bind and are converted to their products. The shape of this site must be maintained for the enzyme to remain functional.

Active transport Active transport is the mediated transport of biochemicals, and other atomic/molecular substances, across membranes. Unlike passive transport, this process requires chemical energy. In this form of transport, molecules move against either an electrical or concentration gradient (collectively termed an electrochemical gradient). This is achieved by either altering the affinity of the binding site or altering the rate at which the protein changes conformations.

There are two main types: *primary* and *secondary*. In primary transport, energy is directly coupled to the movement of a desired substance across a membrane independent of any other species. Secondary transport concerns the diffusion of one species across a membrane to drive the transport of another.

Primary active transport directly uses energy to transport molecules across a membrane. Most of the enzymes that perform this type of transport are transmembrane ATPases. A primary ATPase universal to all cellular life is the sodium-potassium pump, which helps maintain the cell potential.

In secondary active transport, there is however no direct coupling of ATP; instead, the electrochemical potential difference created by pumping ions out of cells is used. The three main forms of this are uniport, counter-transport (antiport) and co-transport (symport).

In counter-transport two species of an ion or other solutes are pumped in opposite directions across a membrane. One of these species is allowed to flow from high to low concentration which yields the entropic energy to drive the transport of the other solute from a low concentration region to a high one. An example is the sodium-calcium exchanger or antiporter, which allows three sodium ions into the cell to transport one calcium out.

Many cells also possess a calcium ATPase, which can operate at lower intracellular concentrations of calcium and sets the normal or resting concentration of this important second messenger. But the ATPase exports calcium ions more slowly: only 30 per second versus 2000 per second by the exchanger. The exchanger comes into service when the calcium concentration rises steeply or "spikes" and enables rapid recovery. This shows that a single type of ion can be transported by several enzymes, which need not be active all the time (constitutively), but may exist to meet specific, intermittent needs.

Co-transport also uses the flow of one solute species from high to low concentration to move another molecule against its preferred direction of flow. An example is the glucose symporter, which co-transports two sodium ions for every molecule of glucose it imports into the cell.

Acuminate Tapers toward the end into a long point.

Acute Pertaining to symptoms that develop suddenly.

Acute toxicity ability of a single dose of a compound to poison .

Adaptin A protein that binds to membrane receptors and mediates the formation of clathrin-coated vesicles.

Adaptive radiation Rapid evolution of a group of related organisms into unoccupied ecological niches.

ADAS Agricultural Development and Advisory Service

Adaxial Directed toward the stem of a plant; pertaining to the upper surface of a leaf.

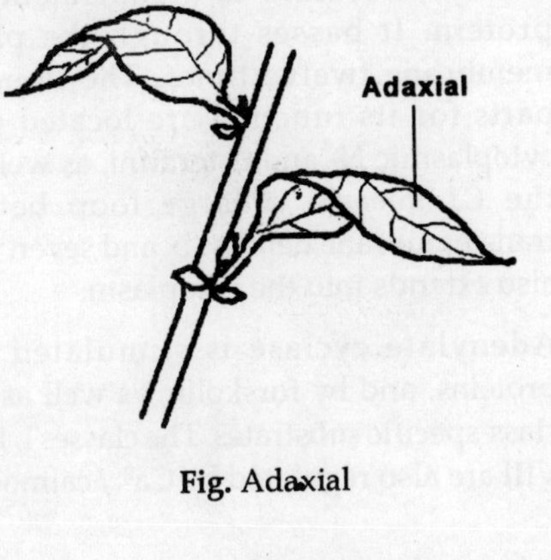

Fig. Adaxial

Addition The movement into or deposition of a quantity of material.

Additive Adjuvant; any of several categories of compounds that can be added to a herbicide solution to enhance its activity; Some examples are oils, surfactants and fertilizers .

Additives Substances added to feedstuffs for preservation or flavour.

Adelphous Means having stamens united into sets; used mostly in combination, as in monoadelphous.

Adenine A purine that base-pairs with either thymine or uracil.

Adenoma A benign tumor arising from glandular epithelium.

Adenylyl cyclase An enzyme that catalyzes the formation of cyclic AMP from ATP.

Adenylate cyclase Adenylate cyclase is a lyase, an enzyme that catalyzes the conversion of ATP to 3',5'-cyclic AMP (cAMP) and pyrophosphate. cAMP is an important molecule in eukaryotic signal transduction, a so-called second messenger. Adenylate cyclase can be activated or inhibited by G proteins, which are coupled to membrane receptors and thus can respond to hormonal or other stimuli. There are nine known adenylate cyclases in mammals.

The reaction that Adenylate Cyclase catalyzes is the conversion of ATP to cAMP. Adenylyl cyclase is a transmembrane protein. It passes through the plasma membrane twelve times. The important parts for its function are located in the cytoplasmic N- and C-termini, as well as in the *C1 domain*, a large loop between transmembrane helices six and seven which also extends into the cytoplasm.

Adenylate cyclase is stimulated by G proteins, and by forskolin, as well as other class-specific substrates. The classes I, III and VIII are also regulated by Ca^{2+}/calmodulin. In neurons, adenylate cyclases are located next to calcium ion channels for faster reaction to Ca^{2+} influx; they are suspected of playing an important role in learning processes. This is supported by the fact that adenylate cyclases are *coincidence detectors,* meaning that they are only activated by several different signals occurring together.

Adherens junction A region of cell-cell adhesion at which the actin cytoskeleton is anchored to the plasma membrane.

Adhesion A molecular attraction between a surface and solvents and solutes in contact with it .

Adjuvant Additive; any of several categories of compounds that can be added to a herbicide solution to enhance its activity; some examples are oils, surfactants and fertilizers .

Adnate In botany, grown together or attached.

Adnation Fusion of unlike parts such as the labellum (of an Orchid) with the column.

Adnexed In botany, annexed or touching, but not attached.

Adobe soil Dense, clay-like soil, also called gumbo, that is also high-alkaline.

Adosculation In botany, the fertilization of plants by the falling of the pollen on the pistils.

Adpressed (appressed) Closely flattened down or pressed against a surface.

Adsorb To accumulate on the soil's surface.

Adsorption The process of a substance binding to a surface.

Adult A fully grown, mature organism

Adventitious In botany, appearing in an abnormal or unusual position or place; occurring as a straggler or away from its natural position, such as roots on aerial stems.

Adventitious buds Buds appearing in an unusual place; for example, buds on leaves.

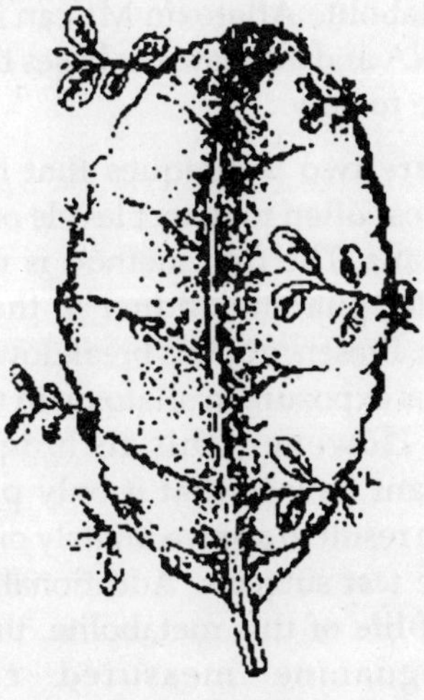

Fig. Adventitious buds

Adventitious root A root of a vascular plant that arises anywhere other than from the radicle or the zone of lateral root formation.

Adventive A plant that has been introduced, but isn't naturalized yet.

Aeciospore Dikaryotic spore of a rust fungus produced in an aecium; in heteroecious rusts, a spore stage that infects the alternate host.

Aecium The fruiting body of a rust fungus in which the first dikaryotic spores (aeciospores) are produced.

Aeration The condition and sum of all processes affecting soil pore-space gaseous composition, particularly with respect to the amount and availability of oxygen for use by soil biota and/or soil chemical oxidation reactions.

Aerator Any implement that is used to break up compacted soil and facilitate air and gas exchange.

Aerator sandals Sandal-type shoes or spikes strapped onto regular shoes used to break up compacted soil in lawns.

Aerenchyma This is spongy tissue full of air pockets that helps leaves and stems of aquatic plants to float on the surface of the water.

Aerial Occurring in the air.

Aerial root This is a root that grows from the stems or shoots of a plant and helps to sustain it by absorbing moisture from the air. Many Orchids and other tropical plants have aerial roots.

Aerification The act of infusing or forcing air into, for example, soil.

Aerobic Condition that has ample amount of oxygen. This condition will allow organisms that require molecular oxygen to growth and function well.

Aerobic digestion Decomposition of organic matter by micro organisms in the presence of oxygen

Aflatoxin Aflatoxins are naturally occurring mycotoxins that are produced by many species of Aspergillus, a fungus, most notably Aspergillus flavus, Aspergillus niger and Aspergillus parasiticus. Aflatoxins are toxic and carcinogenic to animals, including humans. After entering the body, aflatoxins are metabolized by the liver to an intermediate reactive, aflatoxin M_1, an epoxide.

Aspergillus is common and widespread in nature and are most often found when crops are exposed to a high humidity environment over a long period of time or are damaged in stressful conditions such as drought, a condition which lowers the barrier to entry.

The native habitat of Aspergillus is in soil, decaying vegetation, hay, and grains undergoing microbiological deterioration and it invades all types of organic substrates whenever and wherever the conditions are favorable for its growth. Favorable conditions include high moisture content (at least 7%) and high temperature.

Crops which are frequently affected include cereals (maize, sorghum, pearl millet, rice, wheat), oilseeds (peanut, soybean, sunflower, cotton), spices (chile peppers, black pepper, coriander, turmeric, ginger), and tree nuts (almond, pistachio, walnut, coconut).

The toxin can also be found in the milk of animals which are fed contaminated feed. Virtually all sources of commercial peanut butter contain minute quantities of aflatoxin, but it is usually far below the US Food and Drug Administration's (FDA) recommended safe level.

In 1989, Saddam Hussein ordered the government of Iraq to begin production of aflatoxin as an economic biological/ chemical weapon. The methods used by Iraqi scientists at the Salman Pak facility were crude; Aspergillus was grown on wet rice and the final product was reportedly highly impure. According to UNSCOM estimations, 2200 L of aflatoxin were produced. Some of this material was reportedly loaded into missiles, though this allegation has never been proven as no aflatoxin was found after the 2003 invasion of Iraq by the United States. The presumption is that the aflatoxin was destroyed.

High-level aflatoxin exposure produces an acute necrosis, cirrhosis, and carcinoma of the liver exhibited by hemorrhage, acute liver damage, edema, alteration in digestion, and absorption and/or metabolism of nutrients. No animal species is immune to the acute toxic effects of aflatoxins including humans; however, humans have an extraordinarily high tolerance for aflatoxin exposure and rarely succumb to acute aflatoxicosis.

Chronic, subclinical exposure does not lead to as dramatic of symptoms as acute aflatoxicosis. Children, however, are particularly affected by aflatoxin exposure which leads to stunted growth and delayed development. Chronic exposure also leads to a high risk of developing liver cancer, as the metabolite Aflatoxin M1 can intercalate into DNA and alkylate the bases through its epoxide moiety.

There are two techniques that have been used most often to detect levels of aflatoxin in humans. The first method is measuring the AFM_1-guanine adduct in the urine of subjects. Presence of this breakdown product indicates exposure to aflatoxin in the past 24 hours. However, this technique has a significant flaw in that it only produces a positive result in approximately one-third of positive test subjects. Additionally, due to the half-life of this metabolite, the level of AFM1-guanine measured can vary significantly from day to day, based on diet, and thus is not useful for assessing long term exposure.

Another technique, which has been used is a measurement of the AFB_1 - albuminadduct level in the blood serum. This approach is significantly more accurate, as positive results are generated in 90% of positive test subjects. This test is also useful for measuring long-term exposure, as it remains positive for two to three months.

At least 13 different types of aflatoxin are produced in nature. Aflatoxin B_1 is considered the most toxic and is produced by both Aspergillus flavus and Aspergillus parasiticus. Aflatoxin G_1 and G_2 are produced exclusively by A. parasiticus. While the presence of Aspergillus in food products does not always indicate harmful levels of aflatoxin are also present, it does imply a significant risk in consumption of that product.

Aflatoxin B_1 & B_2 : Produced by Aspergillus flavus and A. parasiticus.

Aflatoxin G_1 & G_2 : Produced by Aspergillus parasiticus.

Aflatoxin M_1 : Metabolite of Aflatoxin B_1 in humans and animals (exposure in ng can come from mother's milk).

Aflatoxicol : Studies have shown that concurrent infection with hepatitis B virus (HBV) during aflatoxin exposure increases the risk of hepatocellular carcinoma (HCC). As HBV interferes with the ability of hepatocytes to metabolize aflatoxins, an aflatoxin M_1-DNA conjugate exists for a longer period of time in the liver, increasing the probability of damage to oncogenes such as p53. This effect is synergistic with the resulting damage far greater than just the sum of aflatoxin or HBV individually.

Decreasing HBV infection levels through vaccination is an effective and simple approach that can be taken to reduce these harmful synergistic effects, thus decreasing the impact of chronic aflatoxin exposure. This strategy may prove to be highly effective - many regions of the world which have high aflatoxin rates, such as western Africa and China, also have high HBV infection rates.

AFLP Amplified Fragment Length Polymorphism. A molecular marker method consisting of 1. DNA cleavage with two restriction enzymes,

2. ligation to synthetic linkers, and

3. two PCR amplification steps using primers 16 to 17 bases in length. The technique results in a large number of DNA bands per analysis (50 to 100 are commonly observed), and thus it can be an efficient way of detecting polymorphic loci. However, the large number of bands can be difficult to score. AFLPs are usually dominant and require no prior genome information. Because AFLP bands are "anonymous" (i.e., their chromosome positions are not known in advance), RFLP or SSR markers are sometimes evaluated in the same population in order to "anchor" the AFLPs to specific chromosome arms. AFLPs can be detected either with radioactive label or silver staining.

After ripening A period of dry storage of a seed needed to overcome embryo dormancy

Agarose (or agar) A galactose polymer obtained from the cell walls of some species of red algae or seaweed (Sphaerococcus euchema) and species of Gelidium and Gracilaria, chiefly from eastern Asia, Chile and California. It is also known as Kanten, Agar-Agar, or Agal-Agal (Ceylon Agar). Chemically, agar is a polymer made up of subunits of the sugar galactose; it is a component of the algae's cell walls. Dissolved in hot water and cooled, agar becomes gelatinous; its chief use is as a culture medium for microbiological work. Other uses are as a laxative, a vegetarian gelatin substitute – a thickener for soups, in jellies, ice cream and Japanese desserts such as anmitsu, as a clarifying agent in brewing, and for paper sizing fabrics.

Agar is typically sold as packaged strips of washed and dried seaweed, or in powdered form. Raw agar is white and semi-translucent. For making jelly, it is boiled in water at a concentration of about 0.7-1% w/v (e.g. a 7 gram packet of powder into 1 litre of water would be 0.7%) until the solids dissolve, after which sweeteners, flavouring, colouring, and pieces of fruit may be added. The agar-agar may then be poured into molds or incorporated into other desserts, such as a jelly layer on a cake.

One of the latest fad diets in Asia is the Kanten Diet. Once ingested, kanten triples in size and absorbs water. This results in the consumer feeling more full. Recently this diet has received some press coverage in the United States as well. The diet has shown promise in obesity studies, but agar/kanten has virtually no nutritional value. It is approximately 80% fiber, so part of the diet's effectiveness may be a result of it working as a laxative. There are also some (claimed) effects as to the benefits of agar-agar in controlling diabetes.

Research grade agar is used extensively in plant biology as it is supplemented with a nutrient and vitamin mixture that allows for seedling germination in petri dishes under sterile conditions (given that the seeds are sterilized as well). Nutrient and vitamin supplementation for *Arabidopsis thaliana* is standard across most experiemental conditions. Murashige & Skoog (MS) nutrient mix and Gamborg's B5 vitamin mix are generally used. A 1.0% agar/0.44% MS+vitamin dH20 solution is suitable for growth media between normal growth temps.

It is important to note that the solidification of the agar within any growth media (GM) is pH-dependent, with an optimal range between 5.4-5.7. Usually, the application of KOH is needed to increase the pH to this range. A general guideline would be about 600 µl 0.1M KOH per 250 ml GM. This entire mixture can be autoclaved using the liquid cycling.

This medium nicely lends itself to the application of specific concentrations of phytohormones etc. to induce specific growth patterns in that you can easily prepare a solution contianing the desired amount of hormone, add it to the known volume of GM and autoclave to both sterilize and evaporate off any solvent you mave have used to dissolve the often polar hormones in. This hormone/GM solution can be spread across the surface of petri dishes sown with germinated and/or etiolated seedlings.

Nutrient agar is used throughout the world as a medium for the growth of bacteria and fungi, but not viruses (however, viruses are often grown in bacteria that are growing on agar). Though less than 1% of all existing bacteria can be grown successfully, the basic agar formula can be used to grow most of the microbes, whose needs are known. More specific nutrient agars are available, because microbes can be picky. For example, blood agar, which is generally combined with horse blood, can be used to detect the presence of haemorrhagic micro-organisms such as E.coli O:157 H:7. The bacteria digest the blood, turning the plate clear.

Selective media is agar specially treated to apply a selective pressure to organisms growing on it — for example, to select for salt-tolerant, gram-positive, or gram-negative bacteria. To select for only gram negative organisms you would use MacConkey agar, which would also in turn tell you if the gram negative organism is a lactose fermenter or not indicated by red colonies instead of translucent (non- lactose fermenter).

Differential media includes an indicator that causes visible, easily detectable changes in the appearance of the agar gel or bacterial colonies in a specific group of bacteria. For example, EMB (Eosin Methylene Blue) agar causes E. Coli colonies to have a metallic green sheen, and MSA (Mannitol Salt Agar) turns yellow in the presence of mannitol fermenting bacteria. Hysteresis describes the phenomenon of the differing liquid-solid state transition temperatures that agar exhibits. Agar melts at 85 °C and solidifies from 32-40 °C.

Aggregates Soil particles glued or cemented together. The agents responsible for cementing particles include organic matter and calcium carbonates.

Aggressiveness Relative ability of a plant pathogen to colonize and cause damage to plants .

Agro forestry Cultivated mixtures of trees, crops and/or livestock.

Agrobacteria A soil bacteria that can be a plant pathogen. It works as a natural genetic engineer by invading a plant through wounds and inserting a piece of DNA into the plant's chromosomes. The plant will then begin producing the protein encoded by the inserted gene. Bacillus thuringiensis (Bt) is a strain of agrobacteria.

Agrobacterium A soil bacteria that can be a plant pathogen. It works as a natural genetic engineer by invading a plant through wounds and inserting a piece of DNA into the plant's chromosomes. The plant will then begin producing the protein encoded by the inserted gene. Bacillus thuringiensis (Bt) is a strain of agrobacteria.

Air layering This is a method of propagation in which roots are induced to form around a plant's stem. This is done by removing a very narrow strip of bark from around the branch or stem, about 8 to 12 inches from the tip of the shoot, or at that same distance from the tip, a cut can be made about a third way through the branch in an upward direction and a sliver of wood can be inserted into the cut to keep it open. Next, a bundle of moist sphagnum moss is tied securely around the cut area. The moss must always remain moist (this may be done by wrapping it in polyethylene film) and the roots of the plant somewhat dry. New roots will soon sprout from the incision. The new plant is then cut off below the moss, potted and kept in a humid atmosphere until it is established.

Air plant An epiphyte . Many epiphytic Orchids and Bromeliads are popularly called air plants.

Air pollution Any air contaminant that causes undesirable effects on living organisms or materials.

Air-filled porosity The percent of air occupying the empty spaces of soil and is inversely related to the percent of water occupying the empty spaces of soil. Total porosity= air-filled porosity plus water-filled porosity.

Akinete A resistant resting spore of Cyanobacteria, equivalent to an endospore of Eubacteria.

Akt A protein-serine/threonine kinase that is activated by PIP3 and plays a key role in signaling cell survival.

Alate Winged .

Albino (albinism) White or light-colored; having a marked deficiency in pigmentation.

Albumen In botany, any form of nutritive matter stored within the seed. Also called endosperm.

Aleurone (Aleurone layer) The outermost cell layer of the endosperm, usually only one cell thick in wheat and the only endosperm tissue alive at maturity. The cells of this layer are responsible for the de-novo synthesis of enzymes needed at germination.

Alga (algae) A photosynthetic organism that is not a member of the eubacteria or the Kingdom Plantae.

Algorithm A mathematical equation used to make a prediction, for example, predicting fertilizer requirement from soil analysis and anticipated crop yield.

Alkali soil 1. Highly alkaline soil with a pH of 8.5 or higher.

2. Soluble salt or a mixture of soluble salts present in a soil.

Alkaline Having basic (nonacidic) properties; pH greater than 7.

Alkaline soil Soil with a pH value more than 7, or lime-rich. .

Alkaloid An alkaloid is a nitrogenous organic molecule that has a pharmacological effect on humans and animals. The name derives from the word alkaline; originally, the term was used to describe any nitrogen-containing base (an amine in modern terms). Alkaloids are usually derivatives of amino acids. Most alkaloids have a very bitter taste. They are found as secondary metabolites in plants (e.g., in potatoes and tomatoes), animals (e.g., in shellfish) and fungi, and can be extracted from their sources by treatment with acids (usually hydrochloric acid or sulfuric acid, though organic acids such as maleic acid and citric acid are sometimes used).

While many alkaloids, such as strychnine or coniine, are poisonous, some are used in medicine as analgesics (pain relievers) or anaesthetics, particularly morphine and codeine. Although formally alkaloids, the class of pyrazoles contain two nitrogen atoms in the aromatic ring structure and are not found in nature. They must be produced synthetically.

Alkaloids are usually classified by their common molecular precursors, based on the metabolic pathway used to construct the molecule. When not much was known about the biosynthesis of alkaloids, they were grouped under the names of known compounds, even some non-nitrogenous ones (since those molecules' structures appear in the finished product; the opium alkaloids are sometimes called "phenanthrenes", for example), or by the plants or animals they were isolated from. When more is learned about a certain alkaloid, the grouping is changed to reflect the new knowledge, usually taking the name of a biologically-important amine that stands out in the synthesis process. 1. Pyridine group: piperine, coniine, trigonelline, arecaidine, guvacine, pilocarpine, cytisine, nicotine, sparteine, pelletierine.

2. Pyrrolidine group: hygrine, cuscohygrine, nicotine

3. Tropane group: atropine, cocaine, ecgonine, scopolamine

4. Quinoline group: quinine, quinidine, dihydroquinine, dihydroquinidine, strychnine, brucine, veratrine, cevadine

5. Isoquinoline group: The opium alkaloids (morphine, codeine, thebaine, papaverine, narcotine, sanguinarine, narceine, hydrastine, berberine)

6. Phenethylamine group: mescaline, ephedrine, amphetamine, methamphetamine, MDMA

7. Indole group: a. Tryptamines: DMT, NMT, psilocybin, serotonin, melatonin. b. Ergolines: the ergot alkaloids (ergine, ergotamine, lysergic acid, etc.), LSD. c. Beta-carbolines: harmine, yohimbine, reserpine, emetine

8. Purine group: a. Xanthines: caffeine, theobromine, theophylline

9. Terpenoid group: a. Aconite alkaloids: aconitine. b. Steroids: solanine, samandarin

10. Betaines (quaternary ammonium compounds): muscarine, choline, neurine

11. Pyrazole group: pyrazole, fomepizole

Allele One of the different forms of a gene (or marker) that can exist at a single locus. A single allele for each locus is inherited separately from each parent.

Allelic An allele is one of a pair or series of forms of a gene possible. An example of allelic variation would be a dominant vs recessive allele for one gene.

Allelopathy Ability of one species to inhibit or prevent the growth of another species through the production of toxic substance(s).

Allergen A biological or chemical substance that elicits a reaction involving the body's immune system.

Allochthonous Food material reaching an aquatic community in the form of organic detritus.

Allogamy The process of cross-fertilization in which the pollen of one plant fertilizes the flower of another.

Allosteric regulation The regulation of enzymes by small molecules that bind to a site distinct from the active site, changing the conformation and catalytic activity of the enzyme.

Alluvial soil Referring to the sediment deposited by flowing water.

Alluvium Clay, silt, sand and/or gravel deposited on land by running water.

Alpine Connected with the Alps or any high mountains. Refers to plants growing on mountains above the tree line, which is where the climate becomes too cold for trees to grow.

Alpine house An unheated, well-ventilated greenhouse used for cultivating alpine and bulbous plants.

ALS-AHAS Acetolactate synthase or acetohydroxy acid synthase; an enzyme essential for amino acid production and is the target of many herbicides including the imidazolinones and sulfonylureas; this enzyme has mutated in some species to become herbicide resistant

Alternate (of leaves) Occurring at different levels on opposite sides of a stem. No two leaves are directly across from each other.

Alternation of generations A reproductive cycle in which a haploid phase alternates with a diploid phase.

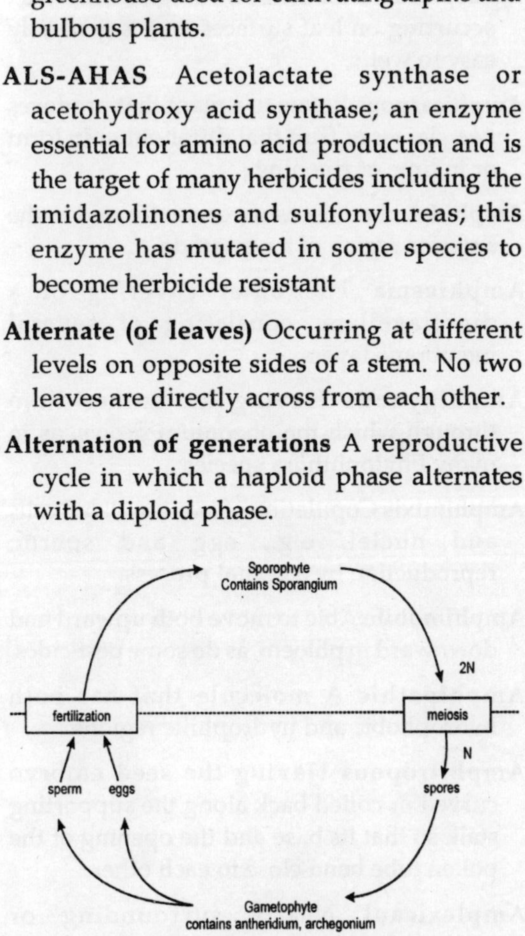

Fig. Alternation of generations

Alternative host A plant other than the main host that a parasite can colonize; alternative hosts are not required for completion of the developmental cycle of the parasite.

Alternative splicing The generation of different mRNAs by varying the pattern of pre-mRNA splicing.

Aluminum A silver-like metallic chemical element that can be found in soil.

Aluminum oxides Group name for the oxides and hydroxides of aluminum. Includes the minerals gibbsite, bauxite.

Aluminum silicates Compounds that contain aluminum, silicon and oxygen.

Aluminum sulfate A colorless salt used as a soil amendment to increase soil acidity.

Amendment Any conditioner or material (ie. Lime, gypsum, etc.) That is added or mixed into the soil to make in more productive.

Ament A catkin; a slim spike-like and usually pendent flower cluster like those found on Willow, Poplar and Birch trees. Aments are either male or female. Some plants produce both male and female aments; others produce male aments on one plant and female on another.

Aminization Process by which microorganisms break down complex proteins to simpler amino acids, amides, and amines; first step of mineralization as part of the nitrogen cycle.

Amino acid Monomeric building blocks of proteins, consisting of a carbon atom bound to a carboxyl group, an amino group, a hydrogen atom, and a distinctive side chain.

Amino acid synthesis inhibitors A mode of action of herbicides that inhibit the production of essential amino acids.

Aminoacyl tRNA synthetase An enzyme that joins a specific amino acid to a tRNA molecule carrying the correct anticodon sequence.

Ammonia 1. A colorless gas made up of one part nitrogen and three parts hydrogenNH3.

2. A gas that can be liquified under pressure to be used as a fertilizer.

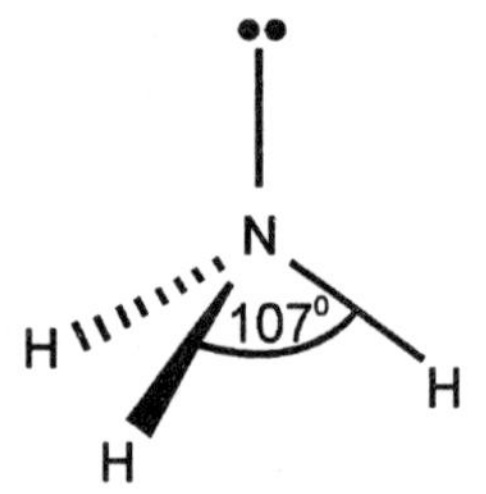

Fig. Ammonia

Ammonia volatilization When ammonia is lost to the atmosphere; occurs when ammonium in the soil, because of pH, is converted to ammonia, which is lost as a gas.

Ammonification The release of ammonium (NH4+)ions from organic matter via microbial action (more broadly termed mineralization)

Ammonium One of the mineral forms of nitrogen. NH4+

Ammonium ion A form that nitrogen takes in many commercial fertilizers.

Ammonium nitrate A dry granule material manufactured by reacting nitric acid with anhydrous ammonia. One-half of the nitrogen in ammonium nitrate is in the nitrate form and one-half is in the ammonium form.

Ammonium nitrate-sulfate A dry nitrogen fertilizer (30-0-0 + 6.5S) manufactured by reacting anhydrous ammonia with a mixture of nitric acid and sulfuric acid; good storage and handling characteristics.

Ammonium phosphate Produced by reacting ammonia with phosphoric acid or a mixture of phosphoric and either sulfuric or nitric acids; the leading nitrogen-phosphorus product in the fertilizer industry; high analysis, high water solubility, good physical characteristics, and low production costs.

Ammonium sulfate An inorganic fertilizer applied to soil (20.6 % N.).

Ammonium thiosulfate solution (ATS)(12-0-0+26S) Most commonly used source of sulfur in fluid fertilizer; weighs 11.5 lbs/gal. After application to the soil, thiosulfate is decomposed to form approximately equal amounts of sulfate sulfur and elemental sulfur.

Amoeboid Having no definite shape to the cell, able to change shape.

Amorphous wax Smooth or glasslike wax occurring on leaf surfaces and is generally easy to wet.

Amphicarpous Refers to a plant that produces two classes of fruit that differ either in form or in time of ripening.

Amphid A chemosensory organ located in the anterior region of a nematode.

Amphiesma The outer covering of a dinoflagellate, consisting of several membrane layers.

Amphigynous Having an antheridium through which the oogonium grows, as in many Phytophthora species .

Amphimixis Copulation of two unrelated cells and nuclei, e.g., egg and sperm; reproduction by a sexual process.

Amphimobile Able to move both upward and downward in phloem, as do some pesticides.

Amphipathic A molecule that has both hydrophobic and hydrophilic regions.

Amphitropous Having the seed embryo curved or coiled back along the supporting stalk so that its base and the opening of the pollen tube bend close to each other.

Amplexicaul Nearly surrounding or embracing the stem, as the base of some leaves.

Amplexifoliate Having leaves that clasp the stem.

Amplification An increase in the number of copies of a specific DNA fragment.

Ampulla A small bladder of flask-shaped organ attached to the roots or immersed leaves of some aquatic plants, as in Nepenthes or Utricularia.

Amyloplast An organelle in some plant cells that stores starch. Amyloplasts are found in starchy plants like tubers and fruits.

Anabolic Synthetic processes resulting in the production of complex products.

Anaerobic Condition that is lacking molecular oxygen. Under this condition, organisms that do not require molecular oxygen will survive and funtion well.

Anaerobic digestion The breaking down of organic matter in oxygen free conditions.

Analogy Non-homologous similarity of structure resulting from similarity of function.

Analysis Need definition.

Anamorph The asexual form in the life cycle of a fungus, when asexual spores (such as conidia) or no spores are produced .

Anaphase The phase of mitosis during which sister chromatids separate and move to opposite poles of the spindle.

Anaphase A The movement of daughter chromosomes toward the spindle poles during mitosis.

Anaphase B The separation of the spindle poles during mitosis.

Anaphase-promoting complex A ubiquitin ligase that triggers progression from metaphase to anaphase by signaling the degradation of cyclin B and cohesins.

Anastomosis Fusion between branches of the same or different structures (e.g. hyphae) to make a network.

Androdioecious Having hermaphrodite (bisexual) flowers on one plant of a species and male flowers on another plant of the same species, but no corresponding form with only female flowers.

Androecium The male organs of a flower; the assemblage of stamens.

Androphore The stalk or column supporting the stamens of certain flowers usually formed by a union of the filaments, as in the Malvaceae and in many genera of Leguminosae.

Anemone centered (Of flowers) Flowers or flower heads in which the central petals or flowers (modified stamens) form a pincushion-like mound and the outer rim or petals or ray florets are flat and spreading, as in some chrysanthemums.

Anemophilous Wind pollinated; refers to flowers that depend on the wind for pollination. Anemophilous flowers are ordinarily unnoticeably small, uncolored and don't secrete honey; however, they produce an abundance of pollen. Examples include the flowers of grasses, pines, sedges, etc.

Angiocarpian A plant that has its fruit enclosed in a husk.

Angiocarpous Having a fruit enclosed within a distinct covering, as the filbert within its husk.

Angiogenesis The formation of new blood vessels.

Angiosperm A plant which flowers and forms fruits with seeds; seeds are enclosed in a pericarp

Angiosperm Any flowering plant that reproduces by seeds enclosed in a protective seed vessel. The term angiosperms is given to the larger of the two divisions of seed-bearing plants, in contrast to gymnosperms, which lack flowers and in which the seeds are naked (Examples are Coniferae, Cycadaceae, etc.)

Angiospermous Having seeds enclosed in a seed vessel, for example, this refers to the poppy, the rose, and most flowering plants, opposed to gymnospermous, or naked-seeded.

Anhydrous ammonia A gaseous material that is compressed and stored as a liquid. At 60oF, a gallon of anhydrous ammonia weights 5.15 pounds. Because anhydrous ammonia needs to go through the nitrification process, it is more resistant to losses from the soil by leaching or denitrification because it is converted by bacterial action to the nitrate form more slowly than other nitrogen sources.

Anhydrous Dry or lacking water.

Anion An ion that is carrying a negative charge of electricity.

Anionic Having a negative charge.

Anisogamous Having flagellated gametes that differ in size and structure.

Anisophyllous Having the leaves of a pair of two different shapes and sizes, as in conifers and many aquatic plants.

Annual A plant that completes an entire life cycle in one growing season. A life cycle includes germination, growth to maturity, blooming, producing seeds and death.

Annual ring In trees and other woody plants, this is the layer of wood produced each year that can be seed when the wood is cut into a cross section. The number of rings equals the age of the tree.

Fig. Annual ring

Annuals Plants that flower and complete thier life cycle in the same year that they are raised from a seed.

Annulations Grooved bands in the cuticle of some nematodes.

Annulus 1. A ring-like part, structure, or membrane.

2. The elastic ring that surrounds the sporecase of most ferns.

3. The raised ring around the center of a succulent flower.

Anoxic No oxygen present.

Antagonism A general term for interference between organisms that may include antibiosis or competition for nutrients or space; action of two or more pesticides that reduces the effectiveness of one or all .

Antagonist A non-phytotoxic chemical that can protect the crop from herbicide injury without reducing the damage imposed on the weed; the mechanism of safenir.g can involve induction of enzymes which metabolize the herbicide in the protected crop; also known as safeners, antidotes or crop protectants.

Antagonize To bother

Antenna A pigment with the primary function of capturing the energy from photons and transferring that energy to other pigments within the photosystem. Most chlorophyll molecules function as antennae, with relatively few of the hundreds of chlorophyll molecules carrying out photochemistry in the reaction center. Carotenoids also function as antennae, but additionally play an important role in quenching triplet chlorophyll and singlet oxygen molecules that would cause damage to the plant cell.

Anterior The underside of a leaf; otherwise called inferior or lower; the opposite of the posterior. Also used in flowers, as on a Snapdragon, where the lower lip is anterior and the upper lip posterior.

Anther The part of a stamen that produces pollen, usually located at the end of a thin stalk in the center of a flower.

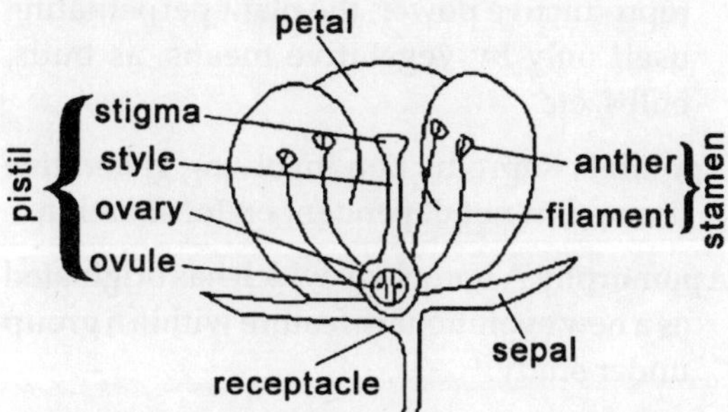

Fig. Anther

Antheridiophore In some liverworts, an elevated structure bearing antheridia.

Antheridium The male reproductive organ in spore-bearing plants such as ferns, mosses, algae, and fungi. Corresponds to the anther in seed-bearing plants.

Antherozoid The thread-like substance produced by the male with which the female organs are fertilized.

Anthers The part of the stamen (part of the male reproductive structure) that develops and holds pollen; supported by the filament

Anthesis The period during which a flower opens or the act of a flower opening; coming to full bloom.

Anthocyanin One of a class of water-soluble pigments usually found in the vacuoles of plant cells; anthocyanins are often blue, pink, or yellow.

Anthography The description of flowers.

Antholeucin The dissolved coloring matter in white flowers.

Anthophore An elongated segment as the base of a flower, between the stem and the base of the petals, stamens, and pistil, as in pinks (Dianthus) or catch-fly (Silene). Also called anthophorum.

Anthoropic soil Soil worked by humans in a way to produce new characteristics that make it different from the original natural soil.

Anthoxanthin The yellow or orange coloring matter of yellow flowers and fruit.

Anthracnose Disease caused by acervuli-forming fungi (order Melanconiales) and characterized by sunken lesions and necrosis.

Anthropogenic The effects of human influence/activities on nature.

Antibiosis An association between organisms, or between an organism and a metabolic product of another organism, that is harmful to one of them.

Antibiotic A chemical compound produced by one microorganism that inhibits growth or kills other living organisms.

Antibody A specific protein formed in the blood of warm-blooded animals in response to the presence of an antigen.

Anticodon The nucleotide sequence of transfer RNA that forms complementary base pairs with a codon sequence on messenger RNA.

Antigen Any foreign chemical (normally a protein) that induces antibody formation in warm-blooded animals.

Antipetalous Having stamens opposite the petals.

Antipodal cells A group of cells, situated at the opposite end to the micropyle, in the mature embryo sac of flowering plants. They are very large and have highly endopolyploid nuclei.

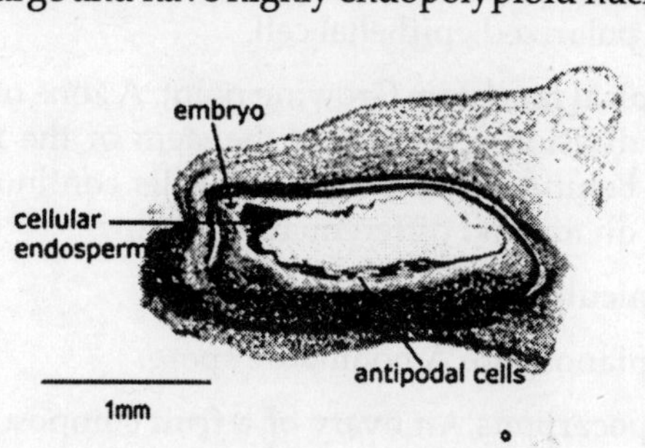

Fig. Antipodal cells

Antiport The transport of two molecules in opposite directions across a membrane.

Antisepalous Standing opposite to sepals; applied to stamens.

Antiserum Blood serum containing antibodies.

Antithesis The time of flower opening in a plant.

AONB Area of Outstanding Natural Beauty.

AP endonuclease A DNA repair enzyme that cleaves next to apyrimidinic or apurinic sites in DNA.

Apatite 1. The main mineral found in phosphate rock.

2. An inorganic compound found in bones; a component of the fertilizer, bone meal.

Aperture Small opening, for example the opening in the test of a foram.

Apetalous Without petals or corolla.

Apex The end of a leaf or growing tip of a shoot.

Apheliotropic Turning away from the light; applied to shoots or other parts of plants, as in roots that turn away from the sun; opposite heliotropic.

Aphid Small, sucking insect of the family Aphididae (order Homoptera) that produces honeydew and injures plants when in large populations.

Aphyllous Without leaves; applied to flowering plants that are naturally leafless, such as many kinds of cactus.

Apical domain The exposed free surface of a polarized epithelial cell.

Apical meristem Growing point. A zone of cell division at the tip of the stem or the root. Behind the meristem the cells continue to divide and differentiation begins.

Apicule A short, sharp point.

Aplanospore A nonmotile spore.

Apocarpous An ovary of a fruit composed of one or more simple and distinct pistils, as in Buttercups.

Apogamous The nature of or characterized by apogamy.

Apogamy Generally, the absence of sexual reproductive power, the plant perpetuating itself only by vegetative means, as buds, bulbs, etc.

Apomixis Reproduction involving generative tissues but not dependent on fertilization.

Apomorphy A homology which has originated as a new evolutionary feature within a group under study.

Apophyllous Having distinct leaves, applied to a whorled flower with distinct sepals and petals; opposed to gamophyllous.

Apoplastic path Water moves in the cell wall or non-living region between cells without crossing any membranes. Apoplastic means 'non-living' tissue.

Apoptosis An active process of programmed cell death, characterized by cleavage of chromosomal DNA, chromatin condensation, and fragmentation of both the nucleus and the cell.

Apothecium Open, cuplike or saucerlike, ascus-bearing fungal fruiting body (ascocarp), often supported on a stalk.

Appendage That which is attached to something as a proper part of it; a subordinate, attached part of anything. Specifically, any subsidiary part added to another part, as hairs and glands to a stem or leaf.

Applanate Flattened out or horizontally expanded.

Appressed Pressed closely against; fitting closely to; lying flat against.

Appressorium Swollen, flattened portion of a fungal filament that adheres to the surface of a higher plant, providing anchorage for invasion by a fungus.

Apterous Wingless.

Aqua ammonia Anyhdrous ammonia dissolved in water. It is a low pressure solution and contains free ammonia, the amount of which increases as air temperature increases.

Aquaculture The cultivation of aquatic organisms for the production of human food. (i.e. Fish farming)

Aquatic Plants that are naturally found growing in water at or above the surface of the soil. There are different types of aquatic plants. Some are emergent or only partially submerged, such as Cattails; some are completely submerged, such as Anacharis; some float with roots anchored to the soil under water, such as Water Lilies; some float freely with roots free in the water, such as water Hyacinths.

Aquifer A water-bearing formation, usually underground, through which water can move easily.

Arabidopsis thaliana A small flowering plant used as a model for plant molecular biology and development.

Arabidopsis The first flowering plant to have its entire genome sequenced. Commonly known as thale cress, a relative of the mustard plant, it has become a model organism in plant biology research because Arabidopsis has almost all of the same genes as other flowering plants and has relatively little "non-coding" DNA.

Arable Able to be cultivated for agriculture.

Arbor A bower formed by trees, vines, or shrubs intertwined or trained over a latticework trellis, so as to make a leafy roof. They are usually provided with seats; any shaded walk. Arbors are used to provide support for plants, to create shade, or to frame a view.

Arboraceous Pertaining to or of the nature of trees.

Arborescent Resembling a tree; tree-like in growth, size, or appearance; having the nature and habits of a tree; branching like a tree.

Arboretum A placė in which trees and shrubs, especially rare ones, are grown for scientific or other purposes; a botanical tree garden.

Arboriculture The cultivation of trees; the art of planting, training, pruning, and cultivating trees and shrubs for decorative purposes.

Arborist A specialist in cultivating trees, in diagnosing tree diseases, performing tree surgery and maintaining trees.

Arbuscle A dwarf tree or tree-like shrub, in size between a tree and a shrub.

Arbuscular mycorrhiza (abbr. AM; endomycorrhiza) Symbiotic association between a nonpathogenic or weakly pathogenic fungus and the roots of plants in which fungal hyphae invade cortical cells of the root and produce vesicles and arbuscles.

Arbuscule Branched haustorial structure of certain endomycorrhizal fungi that forms within living cells of the root.

Archaea Micro-organisms that belong to a major division of life, as different from bacteria as humans are.

Archaebacteria One of two major groups of prokaryotes; many species of archaebacteria live in extreme conditions similar to those prevalent on primitive Earth.

Archegoniophore In some liverworts, an elevated structure bearing archegonia.

Archegonium (archegonia) A multicellular female gametangium found in the embryophytes.

Areole The cushion-like growing point (actually a modified leaf bud) on a cactus from which grows its spines, leaves, side shoots, offsets, or flowers. By identifying the areole on a plant, you can determine if it is a cactus, as only cactus have areoles. The areole generally has some wool or small hairs also growing from it.

ARF A GTP-binding protein required for vesicle budding from the trans-Golgi network.

Aril Applied to the coverings or appendages of seeds.

Aristate Bearded; bristle tipped.

Ark A moveable outdoor pig shelter.

Aromatic amino acids Amino acids whose side chains chemical structures contain an aromatic ring. Examples include phenylalanine, tyrosine and tryphan.

Arthropod Member of the phylum Arthropoda, which consists of animals with articulated bodies and limbs and which includes insects, arachnids, and crustaceans.

Articulate Provided with joints or nodes, at which points separation takes place naturally. Jointed, consisting of segments united by joints. An example is Christmas Cactus (Schlumbergera bridgesii)

Artificial light Light other than sunlight, often fluorescent tubes, used to grow plants.

Ascending Rising somewhat diagonally and curving upward.

Ascidium Any tubular, horn-shaped, or pitcher-like formation of a plant, usually arising from the union of the margins of a leaf or other organ. The ascidium, commonly known as a pitcher, is usually covered with a lid and contains a fluid in which insects are drowned and dissolved. Examples are the pitcher plants (Nepenthes and Sarracenia).

Ascocarp (ascoma) Sexual fruiting body of an ascomycetous fungus that produces asci and ascospores; e.g. apothecium, ascostroma, cleistothecium, perithecium, pseudothecium.

Ascogenous Pertaining to ascus-producing hyphae.

Ascogonium A specialized cell that gives rise to the hyphae that produce asci.

Ascoma Sexual fruiting body of an ascomycetous fungus that produces asci and ascospores; e.g. apothecium, ascostroma, cleistothecium, perithecium, pseudothecium.

Ascomycetes A group of fungi, also called the Ascomycota, that produces sexual spores (ascospores) within a saclike structure called an ascus.

Ascospore Sexual spore borne in an ascus.

Ascostroma A fruiting body containing bitunicate (double-walled) asci in locules (cavities); usually dark with multiple locules, but sometimes single .

Ascus Saclike structure containing ascospores (typically eight) and usually borne in a fungal fruiting body.

Aseptate Descriptive of a hypha or filament lacking cross-walls between the nuclei; ordinarily used to describe fungi.

Asexual Not-sexual; having no sex.

2. Produced by other than sexual processes, as asexual reproduction. Asexual reproduction is any process of propagation that isn't effected by means of sexual organs, as when propagation is carried out by buds, offshoots, bulbs, etc.

Asexual phase The portion of an organisms life cycle devoted to asexual reproduction.

Asexual reproduction Formation of offspring that are genetically identical to the parent, by mitotic division.

Asparagus-knife A tool for prying and pulling out long-rooted plants. Also called dandelion weeder or fishtail weeder.

Aspect The predominant direction of slope of the land.

Asperifoliate Having rough-feeling leaves.

Assimilation The conversion of a substance taken into the living tissues of plants.

Asthenosphere The layer in the Earth below the lithosphere that is weak, probably due to the presence of some molten material. It is this weak layer that allows the tectonic plates to move.

Astral microtubules Microtubules of the mitotic spindle that extend to the cell periphery.

ATP ATP is short for adenosine triphosphate; it is a high-energy molecule used for energy storage by organisms. In plant cells, ATP is produced in the cristae of mitochondria and chloroplasts.

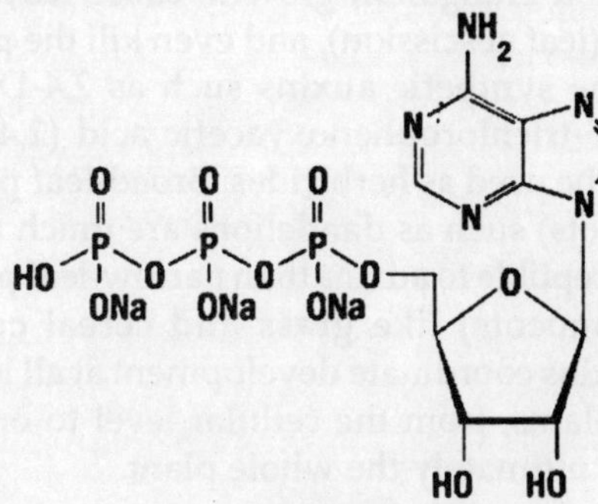

Fig. ATP

ATP synthase A membrane spanning protein complex that couples the energetically favorable transport of protons across a membrane to the synthesis of ATP.

Attar A general term used for a perfume from flowers. In Europe, it denotes only the attar of roses, an essential oil. It takes 150 pounds of roses to provide less than an ounce of attar. The fragrance is too strong when it is concentrated and must be diluted to smell nice.

Attenuate Tapering gradually to a narrow point.

Atypical Having no distinct typical character; not typical; not conformable to the type.

AUDPC (Area Under Disease Progress Curve) A measure of the total amount of disease over a period of time, determined from graphs of disease vs. time, which can be used to compare epidemics quantitatively.

Auger A tool for boring a hole in the soil. Different-sized augers, powered by hand or electric drills, are used in planting or transplanting seeds, seedlings, or bulbs or in fertilizing shrubs, trees and ground covers.

Auricle Ear-shaped projection or appendage. Ear-shaped lobes, often paired, as on the labellum or column of an Orchid.

Auriculate A leaf or petal with a pair of small, blunt, ear-shaped projections at the base.

Austral Relating to or coming from the south, especially the southern hemisphere

Autocarpous Consisting of ripened pericarp alone, having no adnate parts; applied to fruits that are free from the perianth.

Autoclaving A treatment of solutions, tools and containers with high temperature (usually at 120oC) under high pressure (usually at 1.05 kg/cm2) to eliminate organic contaminations.

Autocrine Signaling A type of cell signaling in which a cell produces a growth factor to which it also responds.

Autoecious In reference to rust fungi, producing all spore forms on one species of host plant .

Autogamous Self-fertilized; applied to flowers that are fertilized by their own pollen.

Autogamy Self-fertilization; the fertilization of a flower by its own pollen.

Autophagy The degradation of cytoplasmic proteins and organelles by their enclosure in vesicles from the endoplasmic reticulum that fuse with lysosomes.

Autoradiography The detection of radioisotopically labeled molecules by exposure to X-ray film.

Autotroph An organism that is capable of making and storing food using the sun (or another nonliving) energy sources. Plants are autotrophs because they use water, carbon dioxide, and the energy from the sun to make the complex sugars they use as food.

Autotrophic The ability to synthesize high-energy carbon compounds from inorganic raw materials using energy from sunlight or from certain inorganic chemical reactions.

Auxins A class of plant growth substance (often called phytohormones or plant hormones). Auxins play an essential role in coordination of many growth and behavioral processes in the plant life cycle. Auxins have been demonstrated to be a basic coordinative signal of plant development. Their pattern of active transport through the plant is complex, and auxins typically act in concert with (or opposition to) other plant hormones. For example, the ratio of auxin to cytokinin in certain plant tissues determines initiation of root versus shoot buds. As a result, a plant can (as a whole) react on external conditions and adjust to them, without requiring a nervous system.

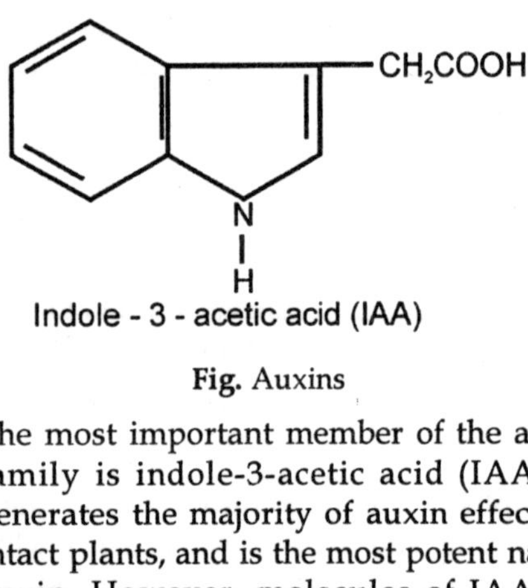

Indole - 3 - acetic acid (IAA)

Fig. Auxins

The most important member of the auxin family is indole-3-acetic acid (IAA). It generates the majority of auxin effects in intact plants, and is the most potent native auxin. However, molecules of IAA are chemically labile in aqueous solution, so IAA cannot be applied commercially as a plant growth regulator.

1. Naturally-occurring auxins include 4-chloro-indoleacetic acid, phenylacetic acid (PAA) and indole-3-butyric acid (IBA).

2. Synthetic auxin analogs include 1-naphthaleneacetic acid (NAA), 2,4-dichlorophenoxyacetic acid (2,4-D), and others.

Auxins are often used to promote initiation of root growth and are the active ingredient of the commercial preparations used in horticulture to root stem cuttings). They can also be used to promote uniform flowering, to promote fruit set, and to prevent premature fruit drop.

Used in high doses, auxin stimulates the production of ethylene. Excess ethylene can inhibit elongation growth, cause leaves to fall (leaf abscission), and even kill the plant. Some synthetic auxins such as 2,4-D and 2,4,5-trichlorophenoxyacetic acid (2,4,5-T) can be used as herbicides. Broad-leaf plants (dicots) such as dandelions are much more susceptible to auxins than narrow-leaf plants (monocots) like grass and cereal crops. Auxins coordinate development at all levels of plants, from the cellular level to organs and ultimately the whole plant.

Although auxins and their effects have been known for a long time, molecular mechanisms of action have been difficult subjects. Auxins directly stimulate or inhibit the expression of specific genes. Auxin induces transcription by targeting for degradation members of the Aux/IAA family of transcriptional repressor proteins, The degradation of the Aux/IAAs leads to the derepression of ARF-mediated transcription. Aux/IAAs are targeted for degradation by ubiquitination, catalysed by an SCF-type ubiquitin-protein ligase.

In 2005, it was demonstrated that the F-box protein TIR1, which is part of the ubiquitin ligase complex SCFTIR1, is an auxin receptor. Upon auxin binding TIR1 recruits specific transcriptional repressors (the Aux/IAA repressors) for ubiquitination by the SCF complex. This marking process leads to the degradation of the repressors by the proteasome, alleviating repression and leading to specific gene expression in response to auxins.

Another protein called ABP1 (Auxin Binding Protein 1) is a putative receptor, but its role is unclear. Electrophysiological experiments with protoplasts and anti-ABP1 antibodies suggest that ABP1 may have a function at the plasma membrane.

On the cellular level, auxin is essential for cell growth, affecting both cell division and cellular expansion. Depending on the specific tissue, auxin may promote axial elongation (as in shoots), lateral expansion (as in root swelling), or isodiametric expansion (as in fruit growth). In some cases (coleoptile growth) auxin-promoted cellular expansion occurs in the absence of cell division. In other cases, auxin-promoted cell division and cell expansion may be closely sequenced within the same tissue (root initiation, fruit growth). In a living plant it appears that that auxins and other plant hormones nearly always interact to determine patterns of plant development.

According to the "acid growth theory," auxins may directly stimulate the early phases of cell elongation by causing responsive cells to actively transport hydrogen ions out of the cell, thus lowering the pH around cells. This acidification of the cell wall region activates enzymes known as expansins, which break bonds in the cell wall structure, making the cell wall less rigid. When the cell wall is degraded (not entirely) by the action of auxins, this now-less-rigid wall is expanded by the pressure coming from within the cell, especially by growing vacuoles. However, the acid growth theory does not by itself account for the increased synthesis and transport of cell wall precursors and secretory activity in the Golgi system that accomplany and sustain auxin-promoted cell expansion.

Growth and division of plant cells together result in growth of tissue, and specific tissue growth contributes to the development of plant organs. Growth of cells contributes to the plant's size, but uneven localized growth produces bending, turning and directionalization of organs, for example, stems turning toward light sources (phototropism), growth of roots in response to gravity (gravitropism), and other tropisms.

As auxins contribute to organ shaping, they are also fundamentally required for proper development of the plant itself. Without hormonal regulation and organization, plants would be merely proliferating heaps of similar cells. Auxin employment begins in the embryo of the plant, where directional distribution of auxin ushers in subsequent growth and development of primary growth poles, then forms buds of future organs. Throughout the plant's life, auxin helps the plant maintain the polarity of growth and recognize where it has its branches (or any organ) connected.

An important principle of plant organization based upon auxin distribution is apical dominance, which means that the auxin produced by the apical bud (or growing tip) diffuses downwards and inhibits the development of ulterior lateral bud growth, which would otherwise compete with the apical tip for light and nutrients. Removing the apical tip and its suppressive hormone allows the lower dormant lateral buds to develop, and the buds between the leaf stalk and stem produce new shoots which compete to become the lead growth. This behavior is used in pruning by horticulturists.

To cause growth in the required domains, it is necessary that auxins be active preferentially in them. Auxins are not synthesized everywhere, but each cell retains the potential ability to do so, and only under specific conditions will auxin synthesis be activated. For that purpose, not only do auxins have to be translocated toward those sites where they are needed but there has to be an established mechanism to detect those sites. Translocation is driven throughout the plant body primarily from peaks of shoots to peaks of roots. For long distances, relocation occurs via the stream of fluid in phloem vessels, but, for short-distance transport, a unique system of coordinated polar transport directly from cell to cell is exploited. This process of polar auxin transport is directional and very strictly regulated. It is based in uneven distribution of auxin efflux carriers on the plasma membrane, which send auxins in the proper direction.

Effects of Auxin 1. Stimulates cell elongation (if gibberellins are also present, the effect is stronger)

2. Stimulates cell division (if cytokinins are also present)

3. Induces formation and organization of phloem (and xylem)

4. Participates in phototropism, geotropism, hydrotropism and other developmental changes

5. Induces new root formation by breaking root apical dominance induced by cytokinins

6. Induces shoot apical dominance

7. Directly stimulates ethylene synthesis (stimulation of ethylene in lateral buds causes inhibition of its growth and potentiation of apical dominance)

8. Inhibits (in low amounts) ethylene formation and transport of precursor

9. Inhibits abscission prior to formation of abscission layer (inhibits senescence of leaves)

10. Induces sugar and mineral accumulation at the site of application

11. Stimulates Flower initiation

12. Is sex determinator

Auxin binding protein (ABP) A 22 kDa protein located in the endoplasmic reticulum which is thought to recognize auxin and act as a receptor at the cell surface.

Auxin Indole-3-acetic acid (IAA); the first plant hormone discovered; a plant hormone involved in many plant growth processes including embryo development, root and flower development, vascular differentiation, stem elongation, apical dominance and tropic response.

Auxinic herbicide The first selective organic herbicides developed. This family of herbicides is described as synthetic auxins, growth regulators with herbicidal action, or herbicides with growth regulatory activity.

Auxospore The zygote of a diatom. Lacking a silica cell wall, an auxospore is able to expand in size.

Auxotonic Determined by growth. Applied to the movements of plants that are the result of growth, rather than outside stimulation.

Availability This term describes the amount of a nutrient in fertilizer or the soil that a plant can immediately absorb; it may be different from the actual amount of the nutrient present.

Available nutrient Nutrients found in the soil that can be taken in by plants at rates and amounts that are significant to the plants' growth.

Available visible radiation Wavelengths between 0.4 and 0.7 μm also referred to as 'usable radiation': equal to c 45% of direct solar radiation.

Available water (Capacity) The amount of water released between in situ field capacity and the permanent wilting point (usually estimated by water content at soil matric potential of -1.5 Mpa). Available water is not the portion of water that can be absorbed by plant roots, which is plant specific.

Average A number that typifies a set of numbers of which it is a function; the usual or ordinary kind or quality.

Avirulence (avr) gene Gene in a pathogen that usually causes a hypersensitive reaction, is associated with active plant defense reactions in a resistant plant, and causes disease in a susceptible plant.

Avirulent (nonpathogenic) Unable to cause disease .

Avoidance Principle of plant disease control in which plants are grown at times or locations where the pathogen is inactive or not present.

Awl-shaped Tapering from the base to a slender and stiff point.

Awn A sharp point or bristle, commonly found on grass seed and flower heads; the bristle-shaped terminal appendage, such as the beard of barley, wheat and many grasses.

Axenic Culture in the absence of living bacteria or other organisms; pure culture.

Axil The angle formed between a leaf's petiole and a stem where an axillary bud develops. The angle formed on an inflorescence, pedicel and rachis (of an Orchid).

Axillary bud A meristem, or bud, located on a stem just above the attachment of a leaf; ordinarily found in seed plants.

Axillary Pertaining to or growing from the axil of plants; produced in the axil.

Axis The stem; the central part or upright support of a plant to which organs or parts are attached.

Axonemal dynein The type of dynein found in cilia and flagella.

Axoneme The fundamental structure of cilia and flagella composed of a central pair of microtubules surrounded by nine microtubule doublets.

Azonal soil 1. A young soil.

2. A soil having very little or no soil profile development (layers not formed).

3. A soil lacking a well-developed horizon.

B

B horizon Horizons that formed below an A, E, or O horizon and are dominated by obliteration of all or much of the original rock structure and show one or more of the following:

1. illuvial concentration of silicate clay, iron, aluminum, humus, carbonates, gypsum, or silica, alone or in combination;

2. evidence of removal of carbonates;

3. residual concentration of sesquioxides;

4. coatings of sesquioxides that make the horizon conspicuously lower in value, higher in chroma, or redder in hue than overlying and underlying horizons without apparent illuviation of iron;

5. alteration that forms silicate clay or liberates oxides or both and that forms granular, blocky, or prismatic structure if volume changes accompany changes in moisture content; or

6. brittleness.

Bacciferous Producing berries.

Bacilliform Shaped like short rods with rounded ends.

Bacillus thuringiensis The soil bacteria which naturally produces the Bt protein toxic to European corn borer (ECB).

Backcross 1. In hybridizing, the process of crossing a first generation hybrid with one of its own parents to produce a better variety.

2. A plant produced by backcrossing.

Backcross breeding After a plant has been transformed, the event must be moved into an elite genetic background. This is done by mating the transformed plant back to elite plants over several generations.

Backcross To cross (mate) an offspring with one of its parents.

Backcrossing A breeding method used to move one or only a few desirable genes from an agronomically poor crop line to an elite line. This is done by crossing a donor parent to an elite line, and crossing offspring with the 'desired gene(s)' back to the elite parent.

Backfill To fill in a planting hole around a plant's roots with a soil mix.

Bacteria (Singular, bacterium). Extremely small, unicellular, prokaryotic micro-organisms that multiply by cell division, occurring in spherical, rodlike, spiral, or curving shapes and found in virtually all environments; some types are important agents in the cycles of nitrogen, carbon, and other matter, while others cause diseases in humans and animals.

Bacterial streaming Large populations of bacteria that exude from the cut surface of infected plant tissue when observed with a microscope.

Bactericide An agent that kills bacteria; specifically, one that kills bacterial plant diseases.

Bacteriocin A protein antibiotic, one or more types of which can be produced and excreted by certain strains of bacteria.

Bacteriophage Virus which infects and destroys a bacterial host. Some phages, however, will incorporate their DNA into that of their host, and remain dormant for an extended period. For this reason, they have become essential tools of genetic engineers.

Bacterium A prokaryotic, microscopic, single-celled organism with a cell wall that increases by binary fission.

Bait Any material used to lure insects, usually to a pesticide. Commonly used to control earwigs and snails.

Baking soada Sodium or potassium bicarbonate. Sometimes mixed with horticultural oils and applied to plants (usually roses) to prevent fungal diseases, such as black spot and powdery mildew.

Balm 1. An oily, aromatic resin or other substance exuding from trees; any aromatic substance secreted from plants.

Balsam An oily, aromatic resinous substance exuding from trees.

Banding 1. Placing fertilizer in continuous narrow bands and then covering it with soil.

2. Encircling part of a plant (usually the trunk) or a portion of a garden with some type of material that traps, kills, or keeps out pests. Examples include sticky materials, poisonous baits and copper stripping.

Banding trees To fasten a barrier around the trunk of a tree to mark the tree for some attention, such as pruning, or to discourage infestation by crawling insects.

Banner 1. The upper petal, usually broad and growing straight up, of some flowers.

2. One of the usually erect petals of the inner upright section of an iris. Also called the standard.

Bar gene Also, called the PAT gene; it is the bacterial gene that encodes the protein phosphinothricin (PAT) which detoxifies the Liberty herbicide molecule resulting in Liberty resistance.

Barb A stiff bristle or hair terminating an awn or prickle, usually slanted downward or backward.

Barbate Bearded; furnished with long, weak hairs.

Bark All tissues outside the vascular cambium of a woody plant, generally including the cork layers, cork cambium (phellogen), and phloem.

Bark ringing This method is used to try to force fruit trees and some other woody plants to flower. A complete ring is cut around the trunk below the lowest branch and another ring is cut right below the first. The bark between the rings is removed and the scar should be covered with grafting wax.

Basal body A structure similar to a centriole that initiates the growth of axonemal microtubules and anchors cilia and flagella to the surface of the cell.

Basal knob (stylet knob) Structure at the base of a nematode stylet.

Basal lamina A sheetlike extracellular matrix that supports epithelial cells and surrounds muscle cells, adipose cells, and peripheral nerves.

Basal leaves Leaves growing at the base or bottom of a plant.

Base A component of nucleotides, the building blocks of DNA. The four bases found in

DNA are represented by A, T, G, and C, which have the ability to base pair.

Base cation Positively charged ions such as magnesium, sodium, potassium, and calcium that increase pH of water (make it less acidic) when released to solution through mineral weathering and exchange reactions.

Base Pair Formed when complementary nucleotides pair by hydrogen bonding. In DNA, the A nucleotide bonds with T, and G bonds with C. Base pairs form the "rungs" of the DNA ladder and the number of base pairs in a strand can be used to describe the length of DNA.

Base-excision repair A mechanism of DNA repair in which single damaged bases are removed from a DNA molecule.

Baseline A measurement, calculation, or location used as a basis for comparison. In terms of ecology, "baseline" refers to how a system functions in its natural state, or before an experiment.

Basement membrane sbasal lamina.

Base-poor soils Soils that have low levels of base cations such as magnesium, sodium, potassium, and calcium.

Basic Having a pH higher than 7, having a lower concentration of H+ ions.

Basidiocarp (basidioma) Sexual fruiting body of a basidiomycetous fungus.

Basidioma Sexual fruiting body of a basidiomycetous fungus.

Basidiomycetes A fungus group, also called the Basidiomycota, characterized by the formation of external basidiospores on basidia.

Basidiospore Haploid (1N) sexual spore produced on a basidium.

Basidium Specialized cell or organ, often club-shaped, in which karyogamy and meiosis occur, followed by production of externally-borne basidiospores (generally four) that are haploid. There are several types of basidia.

Basifixed Attached at the base or lower end.

Basifugal The growth of leaves that develop from the base upward.

Basin irrigation The irrigation of level areas that are surrounded by ridges or levees to retain the water.

Basipetal Downward from the apex toward the base of a shoot; referring to development in the direction of the base so that the apical part is oldest.

Basolateral domain The surface region of a polarized epithelial cell that is in contact with adjacent cells or the extracellular matrix.

Bast A woody fiber obtained from the phloem, or inner bark used to make hemp, jute, ramie, etc.

Bathymetric map A map of the bottom of the ocean, with water depths indicated by contours that join points of equal water depth.

BC 1 generation (Backcross) 1 generation is the offspring resulting from the first cross back to the recurrent parent in the backcross breeding method.

Beaked Having a beak, or something resembling a beak; beak-shaped; rostrate; or ending in a beak-like point.

Beard 1. A crest, tuft, or covering of spreading hairs.

2. The awn or bristle-like appendage upon the chaff of grain and other grasses.

Bearing strength The average laod per unit area required to rupture a supporting soil mass. Measured in units of kg m-2.

Bed An area within a garden or lawn in which plants are grown.

Bedding plant An ornamental plant grown for its flowers or foliage that is suited by habit for growing in beds or masses.

Beetle Any insect of the order Coleoptera characterized by elytra (thickened outer wings), chewing mouth parts, and complete metamorphosis.

Bendiocarb A synthetic insecticide used to control a variety of house and garden pests including earwigs and ants.

Benign tumor A tumor that remains confined to its site of origin.

Benomyl A systemic fungicide used to control a variety of diseases of ornamentals, usually foliar. It is absorbed into the system of the plant to fight off infection. Common name is Benlate.

Benthos The bottom of a sea or lake.

Benzoic acids Family of herbicides that have activity on synthetic auxins belonging to the growth regulators mode of action.

Berry A small, pulpy fruit, which usually contains several seeds (i.e. Huckleberry, mulberry, etc.); any simple fruit in which the inside is fleshy, except the outer skin (i.e. Tomato, banana, grape, etc.); the dry kernel of certain kinds of grain.

Bicellular Two-celled.

Biennial A plant that needs two seasons of growth to produce flowers and fruit. It produces leaves the first year and blossoms, bears fruit, goes to seed and dies the next.

Bifarious Divided into two parts and pointing in two ways, or arranged in two opposite rows, as leaves that grow only on opposite sides of a branch.

Bifid Divided into two equal lobes or parts by a space in the center; forked, like a snake's tongue; divided in two by a deep cleft.

Biflagellate Having two flagella.

Bifoliolate Having two leaflets as in a compound leaf.

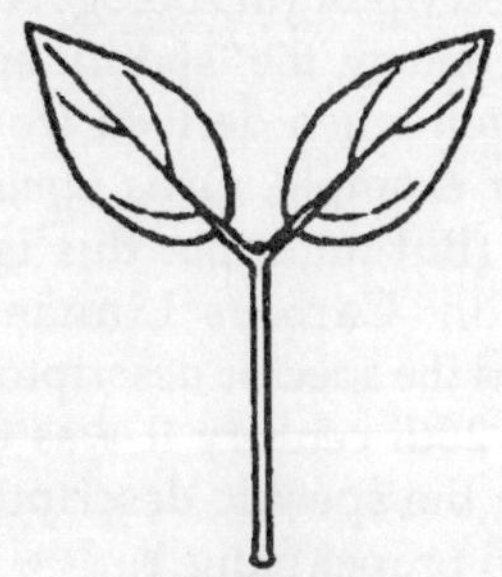

bifoliolate

Fig. Bifoliolate

Bifollicular Having two seed cases, or twin pods, especially in milkweed (Asclepias).

Biforate Having two pores or perforations. Also biforous.

Bifurcate Forked.

Bilabiate Possessing or having the appearance of two lips; the two-lipped center of a flower. (An example is the flowers of Salvia.)

Bilobate Having or divided into two lobes, as a bilobate leaf.

Bilocular Divided into two cell, as in a walnut.

Binary fission A type of asexual reproduction in which two cells, usually of similar size and shape, are formed by the growth and division of one cell.

Binate Being double or in couples; growing in pairs; having only two leaflets to a group.

Binomial nomenclature In biology, binomial nomenclature is the formal method of naming species. As the word "binomial" suggests, the scientific name of a species is formed by the combination of two terms: the genus name and the species descriptor. Although the fine detail will differ, there are certain aspects which are universally adopted: 1. Scientific names are usually printed in italics, such as *Homo sapiens*. When handwritten they should be underlined.

2. The first term (genus name / generic name) is *always* capitalized, while the specific descriptor (in zoology, the "specific name", in botany, the "specific epithet") will not be, even when derived from a proper name. For example, *Canis lupus* or *Anthus hodgsoni*. (But note that this is a modern convention: Carolus Linnaeus always capitalized the specific descriptor, and up to the early 20th century it was common to capitalize the specific descriptor if it was based on a proper name.)

3. When used with a common name, the scientific name usually follows in parentheses, for example, "the house sparrow (*Passer domesticus*) is endangered."

4. The scientific name should generally be written in full when it is first used or when several species from the same genus are being listed or discussed in the same paper or report. It may then be abbreviated by just using an initial (and period) for the genus; for example *Canis lupus* becomes *C. lupus*. In rare cases this abbreviation form has spread to more general use — for example the bacterium *Escherichia coli* is often referred to as just *E. coli*, and *Tyrannosaurus rex* is perhaps even better known simply as *T. rex*.

5. The abbreviation "sp." (or "spec.") is used when the actual specific name is not known: for example *Canis* sp. denotes "a species of the genus *Canis*". The abbreviation "spp." (plural) indicates "several unknown species".

6. The abbreviation "cf." is used when the identification is not confirmed: for example *Corvus cf. splendens* indicates that the species is in doubt.

Binomial nomenclature is also referred to as the 'Binomial Classification System'. The adoption of a system of binomial nomenclature is due to Swedish botanist and physician Carolus Linnaeus (1707–1778) who attempted to describe the entire known natural world and gave every species (mineral, vegetable or animal) a two-part name. However, binomial nomenclature in various forms existed before Linnaeus, and was used by the Bauhins, who lived nearly two hundred years before Linnaeus. Before Linnaeus hardly anybody used binomial nomenclature. After Linnaeus, almost everybody did.

The value of the binomial nomenclature system derives primarily from its economy, its widespread use, and the stability of names it generally favors:

Every species can be unambiguously identified with just two words.

The same name can be used all over the world, in all languages, avoiding difficulties of translation.

Although such stability as exists is far from absolute, the procedures associated with establishing binomial nomenclature tend to favor stability. For example, when species are transferred between genera (as not uncommonly happens as a result of new knowledge), if possible the species descriptor is kept the same. Similarly if what were previously thought to be distinct species are demoted from species to a lower rank, former species names may be retained as infraspecific descriptors.

Despite the rules favoring stability and uniqueness, in practice a single species may have several scientific names in circulation, depending largely on taxonomic point of view.

A major source of instability is the resurrection of forgotten names, which can claim priority of publication. In this case, however, conservation according to the nomenclature Codes is possible.

From the mid nineteenth century onwards it became ever more apparent that a body of rules was necessary to govern scientific names. In the course of time these became Nomenclature Codes governing the naming of animals (*ICZN*), plants (incl. Fungi, cyanobacteria) (*ICBN*), bacteria (*ICNB*) and viruses. These Codes differ. 1. For example, the *ICBN*, the plant *Code* does not allow tautonyms, whereas the *ICZN*, the animal *Code* does allow tautonymy.

2. The starting points, the time from which these *Codes* are in effect (retroactively), vary from group to group. In botany the starting point will often be in 1753, in zoology in 1758. Bacteriology started anew, with a starting point in 1980.

A *BioCode* has been suggested to replace several codes, although implementation is not in sight. There also is debate concerning development of a *PhyloCode* to name clades of phylogenetic trees, rather than taxa. Proponents of the *PhyloCode* use the name "Linnaean Codes" for the joint existing *Codes* and "Linnaean taxonomy" for the scientific classification that uses these existing *Codes*.

The genus name and species descriptor may come from any source whatsoever. Often they are Latin words, but they may also come from Ancient Greek, from a place, from a person (preferably a naturalist), a name from a local language, etc. In fact, taxonomists come up with specific descriptors from a variety of sources, including inside-jokes and puns.

However, names are always treated grammatically as if they were a Latin sentence. For this reason the name of a species is sometimes called its "Latin name," although this terminology is frowned upon by biologists (and philologists), who prefer the phrase scientific name.

Binomial, Latin The scientific name of an organism, composed of two names, the first designating the genus and second the specific epithet, together making the species name.

Binucleate Having two nuclei.

Bioactive molecule Molecules that have a direct or measurable effect on living tissues and cells.

Bioassay Any test (assay) using a living organism.

Biochemical oxygen demand (BOD) Biochemical (biological) oxygen demand is a test used to measure the concentration of biodegradable organic matter present in a sample of water. It can be used to infer the general quality of the water and its degree of pollution and it is used in water quality management and assessment, Ecology and environmental science. BOD is not an accurate quantitative test and should be considered as providing an indicator of the quality of a water body.

Most pristine rivers will have a 5 day BOD of less than 1 mg/l. Moderately polluted rivers may have BODs in the range 2 mg/l to 8 mg/l. Efficiently treated municipal sewage treated by a three stage process would have a BOD value of about 20 mg/l. Untreated sewage is very variable but might average out at around 600 mg/l with slurry from dairy farms at around 2000 mg/l and silage liquor around 6000 mg/l. The influent coming into most waste water treatment plants is about 200 mg/L.

BOD measures the rate of uptake of oxygen by micro-organisms in the sample of water at a fixed temperature and over a given period of time. To ensure that all other conditions are equal, a very small amount of micro-organism seed is added to each sample being tested. This seed is typically generated by diluting activated sludge with de-ionised water. The test generally takes place over an elapsed period of 5 days but other BOD tests are also used.

The BOD test is carried out by diluting the sample with de-ionised water saturated with oxygen , innoculating it with a fixed aliquot of seed, sealing the sample (to prevent further oxygen dissolving in), and then placing it in the dark (to prevent photosynthesis and thereby the addition of oxygen). The sample is kept at 20 °C in the dark and tested for dissolved oxygen (DO) after five days. The apparent BOD for the control is subtracted from the control result to provide the corrected value.

The loss of dissolved oxygen in the sample, once corrections have been made for the degree of dilution, is called the BOD_5. In the UK allylthiourea is also added at the start of the test to prevent oxidation of ammonia. Results from such tests are represented as BOT5(ATU) and referred to as Carbonaceous BOD (CBOD) in the U.S.. Less frequently used is the Ultimate BOD (UBOD) test, in which DO is repeatedly measured by DO meter in the same specialized bottles until it has reached equilibrium.

BOD is similar in function to chemical oxygen demand (COD), in that both measure the amount of organic compounds in water. However, COD is less specific since it measures everything that can be chemically oxidised rather than just levels of biology active organic matter. BOD is used as a gauge of the effectiveness of wastewater treatment plants. Various commercial devices are available for its determination.

The Royal Commission on River Pollution which was established in 1865 and the formation of the Royal Commission on Sewage Disposal in 1898 led to the selection in 1908 of BOD_5 as the definitive test for organic pollution of rivers. Five days was chosen as an appropriate test period because this is supposedly the longest time that river water takes to travel from source to estuary in the UK. In 1912 the commission also set a standard of 30 ppm BOD_5 as the maximum concentration of permitted in sewage works discharging to rivers provided that there was at least an 8:1 dilution available at dry weather flow. This was contained in the famous 30:20 + full nitrification standard which was used as a yardstick in the UK up to the 1970s for sewage works effluent quality.

Biocide A compound toxic to all forms of life.

Biocontrol (biological control) Exploitation by humans of the natural competition, parasitism and/or antagonism of organisms for management of pests and pathogens.

Biodiversity The Term Biodiversity was coined in 1985, (abbreviating 'biological diversity') and has been defined in many different ways. Holdgate (1991) regarded it as the 'Total sum of life's variety on Earth, expressed at the genetic, species and ecosystem level'

Biodyanmic Having to do with the dynamic relation between organisms and their environments.

Biofuel Any fuel that is derived from biomass — recently living organisms or their metabolic byproducts, such as manure from cows. It is a renewable energy source, unlike other natural resources such as petroleum, coal and nuclear fuels. One definition of *biofuel* is any fuel with an 80% minimum content by volume of materials derived from living organisms harvested within the ten years preceding its manufacture.

Like coal and petroleum, biomass is a form of stored solar energy. The energy of the sun is "captured" through the process of photosynthesis in growing plants. Like all methods used to generate energy, the combustion of biomass generates pollution as a by-product. One advantage of biofuel in comparison to most other fuel types is that the energy within the biomass can be stored for an indefinite time-period and without any danger.

Agricultural products specifically grown for use as biofuels include corn and soybeans, primarily in the United States, as well as flaxseed and rapeseed, primarily in Europe, and hemp is a growing crop around the world except for in America. Biodegradable outputs from industry, agriculture, forestry, and households can also be used to produce bioenergy; examples include straw, lumber, manure, sewage, garbage and food leftovers. These feedstocks are converted into biogas through anaerobic digestion. Biomass used as fuel often consists of underutilized types, like chaff and animal waste. Much research is currently in progress into the utilization of microalgae as an energy source, with applications being developed for biodiesel, ethanol, methanol, methane, and even hydrogen. On the rise is use of hemp, although current politics currently restrains it.

Paradoxically, in some industrialized countries like Germany, food is cheaper than fuel compared by price per joule. Central heating units supplied by food grade wheat or maize are available.

Biofuel can be used both for central- and decentralized production of electricity and heat. As of 2005, bioenergy covers approximately 15% of the world's energy consumption. Most bioenergy is consumed in developing countries and is used for direct heating, as opposed to electricity production. However, Sweden and Finland supply 17% and 19% respectively, of their energy needs with bioenergy, a high figure for industrialized countries.

The production of biofuels to replace oil and natural gas is in active development, focusing on the use of cheap organic matter (usually cellulose, agricultural and sewage waste) in the efficient production of liquid and gas biofuels which yield high net energy gain. The carbon in biofuels was recently extracted from atmospheric carbon dioxide by growing plants, so burning it does not result in a net increase of carbon dioxide in the Earth's atmosphere. As a result, biofuels are seen by many as a way to reduce the amount of carbon dioxide released into the atmosphere by using them to replace non-renewable sources of energy. Noticeable is the fact that the quality of timber or grassy biomass does not have a direct impact on its value as an energy-source.

Dried compressed peat is also sometimes considered a biofuel. However, it does not meet the criteria of being a renewable form of energy, or of the carbon being recently absorbed from atmospheric carbon dioxide by growing plants. Though more recent than petroleum or coal, on the time scale of human industrialisation, peat is a fossil fuel and burning it does contribute to atmospheric CO_2.

Biofuel was used since the early days of the car industry. Otto Von Nicklous, the inventor of explosion engine, conceived his invention to run on ethanol. While Rudolf Diesel, the inventor of the combustion engine, conceived it to run on peanut oil. The Ford Model T, a car produced between 1903 and 1926 used ethanol. However, when crude oil began being cheaply extracted from deeper in the soil (thanks to drilling starting in the middle of the 19th century), cars began using fuels from oil. Then with the oil shock of 1973 and 1979, there was an increase interests from governments and academics in biofuels. However, interest decreased with the counter-shock of 1986 that made oil prices cheaper again. But since about 2000 with rising oil prices, concerns over the potential oil peak, greenhouse gas emissions, and stability in the Middle East are pushing renewed interest in biofuels. Government officials have made statements and given aid in favour of biofuels.

Certain types of biomass have attracted research and industrial attention. Many of these are considered to be potentially useful for energy or for the production of bio-based products. Most of these are available in very large quantities and have low market value.

Algae

Bagasse from Sugarcane

Dried distiller's grain

Firewood

Hemp

Jatropha

Landscaping waste

Maiden Grass

Maize(corn)

Manure

Meat and bone meal

Miscanthus

Peat

Plate waste

Rice hulls

Silage

Stover

Switchgrass

Whey

Biologically produced alcohols, most commonly ethanol and methanol, and less commonly propanol and butanol produced by the action of bacteria — see alcohol fuel.

Methanol, which is currently produced from natural gas, can also be produced from biomass — although this is not economically viable at present. The methanol economy is an interesting alternative to the hydrogen economy.

Biomass to liquid, synthetic fuels produced from syngas. Syngas in turn, is produced from biomass by gasification.

Ethanol fuel produced from sugar cane is being used as automotive fuel in Brazil. Ethanol produced from corn is being used as a gasoline additive (oxygenator) in the United States. Cellulostic ethanol is being manufactured from straw (an agricultural waste product) by Iogen Corporation of Ontario, Canada.

Butanol is formed by A.B.E. fermentation (Acetone, Butanol, Ethanol) and experimental modifications of the ABE process show potentially high net energy gains with butanol being the only liquid product. Butanol can be burned "straight" in existing gasoline engines (without modification to the engine or car), produces more energy and is less corrosive and less water soluble than ethanol, and can be distributed via existing infrastructures.

Mixed Alcohols (e.g., mixture of ethanol, propanol, butanol, pentanol, hexanol and heptanol, such as Ecalene™), obtained either by biomass-to-liquid technology (namely gasification to produce syngas followed by catalytic synthesis) or by the MixAlco Process.

Biologically produced oils can be used in diesel engines:

Straight vegetable oil (SVO).

Waste vegetable oil (WVO) - waste cooking oils and greases produced in quantity mostly by commercial kitchens

Biodiesel obtained from transesterification of animal fats and vegetable oil, directly usable in petroleum diesel engines.

Biogas is produced by anaerobic decomposed organic material by anaerobes. Biogas can be produced either from waste materials or by the use of energy crops fed into anaerobic digesters.

Biogas contains methane and can be recovered in industrial systems such as anaerobic digesters and mechanical biological treatment systems. Landfill gas is a less clean form of biogas which is produced in landfills through naturally occurring anaerobic digestion. Paradoxically if this gas is allowed to escape into the atmosphere it is a potent greenhouse gas.

Biologically produced oils and gases can be produced from various wastes:

Thermal depolymerization of waste can extract methane and other oils similar to petroleum.

Pyrolysis oil may be produced out of biomass, wood waste etc. using heat only in the flash pyrolysis process. The oil has to be treated before using in conventional fuel systems or internal combustion engines (water + pH).

One company, GreenFuel Technologies Corporation, has developed a patented bioreactor system that utilizes nontoxic photosynthetic algae to take in smokestacks flue gases and produce biofuels such as biodiesel, biogas and a dry fuel comparable to coal.

One widespread use of biofuels is in home cooking and heating. Typical fuels for this are wood, charcoal or dried dung. The biofuel may be burned on an open fireplace or in a special stove. The efficiency of this process may vary widely, from 10% for a well made fire (even less if the fire is not made carefully) up to 40% for a custom designed charcoal stove. Inefficient use of fuel may be a minor cause of deforestation (though this is negligible compared to deliberate destruction to clear land for agricultural use) but more importantly it means that more work has to be put into gathering fuel, thus the quality of cooking stoves has a direct influence on the viability of biofuels.

American homeowners are turning to burning corn in special stoves to reduce their energy bills. Sales of corn-burning stoves have tripled this year [...] Corn-generated heat costs less than a fifth of the current rate for propane and about a third of electrical heat".

The methane in biogas is often pure enough to pass directly through gas engines to generate green energy. Anaerobic digesters or biogas powerplants convert this renewable energy source into electricity. This can either be used either commercially or on a local scale.

In Germany small scale use of biofuel is still a domain of agricultural farms. It is an official aim of the German government to use the entire potential of 200,000 farms for the production of biofuel and bioenergy.

Different combustion-engines are being produced for very low prices lately. They allow the private house-owner to utilize low amounts of "weak" compression of methane to generate electrical and thermal power (almost) sufficient for a well insulated residential home.

Unfortunately, much cooking with biofuels is done indoors, without efficient ventilation, and using fuels such as dung causes airborne pollution. This can be a serious health hazard; 1.5 million deaths were attributed to this cause by the World Health Organisation as of 2000. There are various responses to this, such as improved stoves, including those with inbuilt flues and switching to alternative fuel sources. Most of these responses have difficulties. One is that fuels are expensive and easily damaged. Another is that alternative fuels tend to be more expensive, but the people who rely on biofuels often do so precisely because they cannot afford alternatives. Organisations such as Intermediate Technology Development Group work to make improved facilities for biofuel use and better alternatives accessible to those who cannot currently get them. This work is done through improving ventilation, switching to different uses of biomass such as the creation of biogas from solid biomatter, or switching to other alternatives such as micro-hydro power.

Direct biofuels are biofuels that can be used in existing unmodified petroleum engines, like biodiesel and biobutanol. Another direct biofuel is E10 (in the US) - however, most cars, even those built prior to 1988, will run on E5 and apparently E20 can be used in most post-1988 American cars without modification. Other than biodiesel, biobutanol and bioethanol, few direct biofuels, if any, exist.

On the other hand, recognizing the importance of bioenergy and its implementation, there are international organizations such as IEA Bioenergy, established in 1978 by the International Energy Agency (IEA), with the aim of improving cooperation and information exchange between countries that have national programs in bioenergy research, development and deployment.

Biogeochemistry The study of the relationship between the geochemistry of a region and the animal and plant life in that region.

Bioinformatics The use of computers to collect, analyze and store genomics information.

Biolistic transformation A method used for genetic engineering in which plant cells are bombarded with metal particles coated with foreign genes using a "gene gun".

Biolistics A method for introducing DNA into a target organism using high-speed particle bombardment.

Biological efficiency The efficiency of an organism or biological system

Biomass The dry weight of organic matter (such as trees, animals, herbaceous plants, etc) in an ecosystem.

Biomolecules Molecules, usually organic molecules, found in cells and living organisms

Biota The plant and animal life of a region or period.

Biotechnology The application of the study of living things to a myriad of processes, such as agricultural production, hybrid plant development, environmental research, and much more.

Biotic factors Biological aspects of the environment, one of the conventionally accepted soil forming factors.

Biotic Pertaining to life and therefore living organisms. For example, plant diseases of a biotic origin are caused by living organisms such as insects, nematodes, etc.

Biotransformation Change in the herbicide's chemical structure within plant cells.

Biotroph (obligate parasite) An organism that can live and multiply only on another living organism .

Biotype A subdivisicn of a species, subspecies, or race based on some identifiable physiological trait such as a specific virulence pattern.

Bipinnate Twice pinnate; doubly divided leaf.

Biplicate Doubly folded; twice folded together, transversely, as the first new leaves that sprout from a bean.

Biramous Possessing two branches; dividing into two branches.

Bird netting Nylon or plastic mesh used as a drape to keep birds out of fruit trees, berry patches, or vegetable gardens. Smaller mesh netting can be placed right over strawberries, raspberries, and corn; larger mesh netting may be draped over trees, bushes, grapes, or over a supporting structure.

Birimose Opening by two slits, as the anthers of most plants.

Bisetose Having two bristles.

Bisexual A flower containing both male and female organs.

Bispinous Having two spines.

Bitunicate Having a double ascus wall .

Black plastic Sheet plastic that is placed around plants as a mulch to obstruct weeds, hold moisture, and absorb heat. Also called black polyethylene or black poly sheeting.

Black smoker A chimney-like structure on the seafloor made of metal sulfides, out of which hot (~350°C) fluids that look like black smoke flow. The black color of the fluid is due to mineral particles within it

Blade A plant's leaf, such as a blade of grass; the expanded part of a leaf, petal, sepal, etc. as distinguished from the petiole or leafstalk.

Blanch To whiten or prevent from becoming green by excluding light. Blanching is applied to the stems or leaves of plants such as celery, lettuce, and endive. This is done by banking up the soil around the stems of the plants, tying the leaves together to keep the inner ones from light, or covering with pots, boxes, etc.

Blastema The point of growth from which any organs or part of an organ is developed.

Blasting A symptom of plant disease characterized by shedding of unopened buds; classically, the failure to produce fruit or seed.

Bleeding When sap or latex oozes from a plant or tree.

Blight A disease characterised by rapid and widespread death of plant tissue. Blight may take the form of extensive spotting, discoloration, wilting, or destruction of leaves, flowers and stems.

Bloom 1. The flower of a plant.

2. The state of blossoming.

3. A powdery, wax-like deposit or coating that covers certain fruits (such as apples, grapes and plums), vegetable leaves (such as cabbage), and succculent leaves.

Blossom 1. The flower of a plant, usually noticeable because of the colored leaflets that form it.

2. To put forth blossoms or flowers; to come into bloom.

Blotch Necrotic area of tissue irregular in form.

Blowball The downy heads of seeds produced after the flowers have died. (i.e. dandelion, salsify, etc.)

Blue Baby Syndrome An illness that begins when large amounts of nitrates in water are ingested by an infant and converted to nitrite by the digestive system. The nitrite then reacts with oxyhemoglobin (the oxygen-carrying blood protein) to form metheglobin, which cannot carry oxygen. If a large enough amount of metheglobin is formed in the blood, body tissues may be deprived of oxygen, causing the infant to develop a blue coloration of their mucous membranes and possibly digestive and respiratory problems. This condition is also known as methemoglobinemia.

Bog garden A garden devoted to moisture-loving plants. Low-lying, marshy sites are good for this.

Bole The trunk of a tree.

Bolting Flower stalk elongation.

Bone meal Ground bone, raw or steamed, that is used as a fertilizer to add phosphoric acid to the soil. It is especially beneficial in herb gardens.

Bonsai A shrub or tree grown in a container and dwarfed by pruning, pinching, and wiring to produce a desired shape.

Bonsai dish A planter for holding a miniaturized tree or shrub.

Border A narrow strip of garden that encloses or separates a portion of lawn or other landform by a path, walk, driveway, street, or building. Usually planted with perennials or flowering shrubs for spring and summer bloom.

Border fence A short fence, usually only 12 to 18 inches high, of wood or metal, used to mark and separate flower beds from foot traffic and lawn mowers.

Boreal Referring to forest areas, found in northern temperate areas, that are dominated by coniferous trees such as spruce, fir, and pine.

Borer Insect or insect larva that forms tunnels or cavities in the bark or within the wood of trees.

Boric acid A salt that can be effectively used as a barrier against some crawling household pests such as ants. If kept dry, it can be used outdoors.

Boron A trace element usually present in most soils but sometimes added by applying fertilizer.

Bosket A grove; a thicket or small plantation in a garden, park, etc., formed of trees, shrubs, or tall plants.

Botanic Pertaining to botany, or the scientific study of plants.

Botanical insecticide Any natural insecticide derived from plants. They include pyrethrum, neem, rotenone, rynia, sabadilla, and nicotine sulfate. They break down quickly when exposed to sun and must be applied often. They are usually non-toxic, but not always.

Botany The science of plants, their classification and study.

Botryoid Having the form of a bunch of grapes.

Bough An arm or branch of a tree.

Boundary layer A region of still air near the surface of a leaf through which water must diffuse en route to the atmosphere

Bow saw A tool having a straight blade attached at both ends to a curved handle, in all making the shape of the letter D. It is used for cutting branches.

Bract A modified leaf at the base of a flower or flower cluster. It may resemble a normal leaf, be reduced and scale-like in appearance, or large and brilliantly colored.

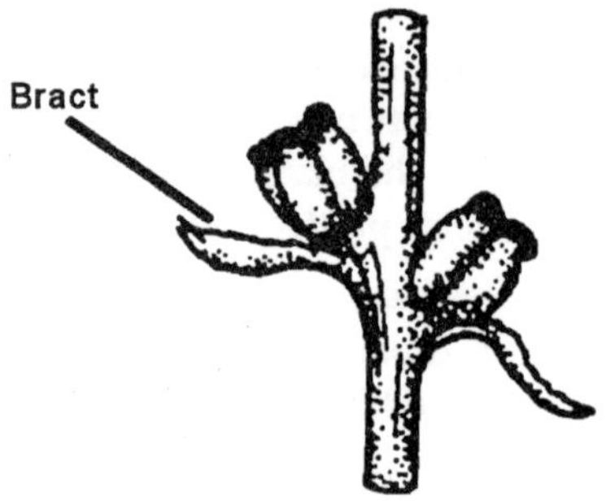

Fig. Bract

Bracteate Having bracts.

Branch A stem growing from the trunk of a tree, shrub, or other plant (the smaller offshoots are called branchlets, twigs, or shoots); a bough.

Branched chain amino acids Amino acids whose side chain chemical structrue is branched. Examples include leucine, isoleucine and valine.

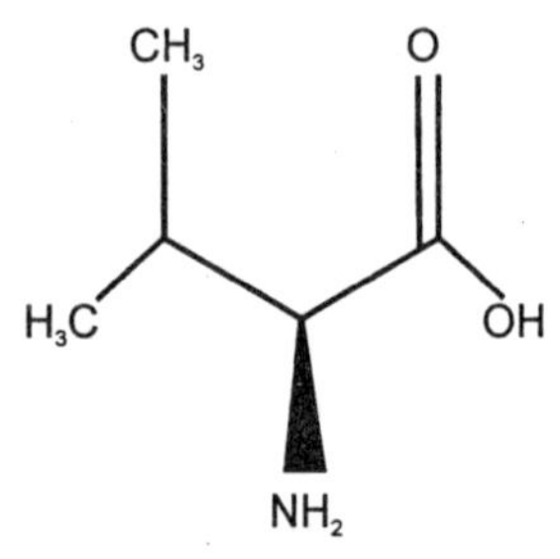

Fig. Branched chain amino acids

Brassicas Plants of the brassica classification e.g. turnip.

Break crop A change in crop in an arable rotation in order to help control weeds and plant diseases, i.e Oilseed rape grown in a sequence of cereal crops would be a break crop.

Breaking Disease symptom, usually caused by a virus, involving addition or loss of flower color to create a variegated pattern. .

Breed A group of plants, developed by man, which will not keep their characteristics in the wild.

Breeding The directed development of better crops through the selection of individuals expressing the desired traits. Breeding alters the genetic makeup of the population and this may be considered a form of genetic modification of that population. The alteration of plant characteristics through breeding was accomplished much earlier than the discovery of modern genetics.

Breeding line Plant strain used in a plant breeding program and usually containing one or more desirable agronomic or breeding characteristics.

Bristle A stiff, sharp hair on a plant.

Broad-base terrace A low embankment that is constructed across a slope to reduce runoff and/or erosion.

Broadcast Scattered upon the ground with the hand, as in sowing seed, instead of sowing in drills or rows.

Broadcast application Application by spreading or scattering on the soil surface.

Brooming Profuse branching of woody stems from single stem position.

Brown forest soil A soil rich in humus with a dark brown surface soil.

Brown podzolic soil A mixed humus and mineral soil with a thin mat of decayed leaves; usually found under a deciduous (leafy) or mixed deciduous and coniferous forest.

Brown rot (of wood) Wood decay resulting from selective removal of cellulose and hemicellulose, leaving a brown amorphous residue that usually cracks into cubical blocks and consists largely of slightly modified lignin.

Brown soil A brown surface soil found under grasses and shrubs that gets lighter in color the deeper you dig.

Browsing A method of feeding by herbivores, in which the leaves and peripheral shoots are removed from trees and shrubs.

Brush The small trees and shrubs of a wood; a thicket or small trees; scrub.

Bt corn Corn that has been transformed with the Bt gene and is resistant to European corn borer.

Bt corn hybrids Corn hybrids that have been transformed with the Bt gene and are resistant to European corn borer (ECB).

Bt gene A gene originating from the Bacillus thuringiensis soil bacteria that encodes a protein toxic to the European corn borer (ECB).

Bt protein Another name for the CryIA proteins. These proteins occur naturally in Bacillus thuringiensis bacteria and are toxic to the European corn borer (ECB).

Bud 1. A growing point enclosed by closely overlaid rudimentary leaves. Buds may contain foliage or flowers. Bulbs and bulbils are forms of leaf buds. Flower buds are unexpanded blossoms.

2. To graft a bud of 1 plant on the stem of another (i.e. to bud a garden rose on a brier).

Bud mutation A bud variation resulting from local genetic alteration and producing a permanent modification that usually can be retained by grafting.

Bud scale A specialized protective leaf of a bud.

Bud sport The product of bud mutation or bud variation.

Bud variation In the outgrowth of a bud, the deviation in any respect from the ordinary growth of the plant, producing what is commonly known as a sport. Many varieties arise this way, such as Red Delicious Apples, and these are perpetuated by bud propagation.

Bud wood Wood consisting of strong, young shoots bearing buds suitable for use in budding.

Budding 1. The production of buds.

2. A form of propagation in which a single bud is grafted to the stock. Also called bud grafting.

Budding strip A rubber strip, typically 1/8th to 3/8th inches wide and up to 8 inches long, used to hold grafts.

Buffer A substance in the soil that will chemically act to resist changes in the soil's reaction or pH, usually clay or fine organic matter.

Buffer pH Test used to estimate lime requirement – amount of agricultural lime required to neutralize the H+ from the soil exchange sites and in the soil solution.

Buffer strips Vegetated areas designed so that runoff passes through the strips for filtration, sedimentation and sorption of nutrients.

Bug Any insect of the order Hemiptera characterized in part by piercing-sucking mouth parts, a triangular scutellum, two pairs of wings, and gradual metamorphosis.

Bulb A short, flattened, usually globose or disc-shaped, underground, perennial, storage organ composed of concentric layers of overlapping fleshy scale leaves attached to a stem plate at the base; essentially a subterranean bud.

Bulb planter A sharp-edged, tapered cylinder used to remove a plug of soil or sod, in which a bulb is placed. The plug is then returned to the hole to cover the bulb.

Bulbils Small bulb or bulb-like structure found in the place of a flower or leaf axil of a plant; found on tiger-lilies

Bulblet A little or secondary bulb; specifically, a small aerial bulb or bud with fleshy scales, growing in the axils of leaves, as in the tiger lily, or taking the place of flower buds, as in the common onion. Also called bulbil.

Bulk density The dry weight of soil per unit volume of soil; most often expressed on a soil volume basis, rather than on a particle basis. Bulk density considers both the solids and the pore space; whereas, particle density considers only the mineral solids.

Bullate Blistered or puckered.

Bunt ball Smut sorus filled, with teliospores, that replaces a cereal or grass kernel but is covered by plant tissue at maturity.

Buoyancy The tendency of a body or fluid to rise when it is less dense than its surroundings.

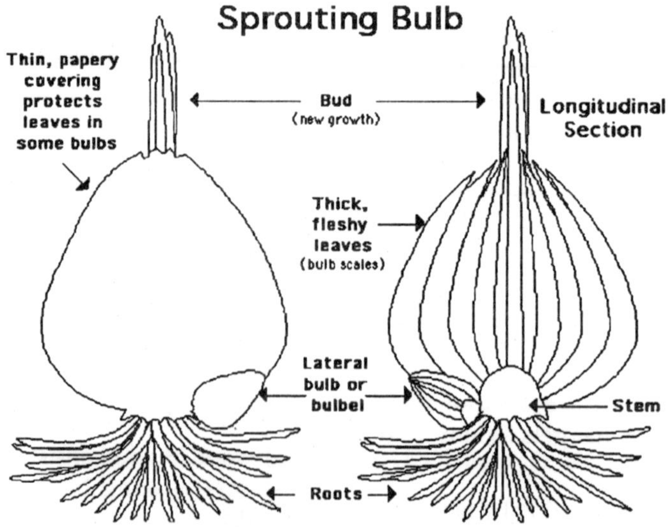

Fig. Bulb

Bur The rough, prickly covering of the seeds of certain plants, such as the chestnut or burdock. Also spelled burr.

Burl A hard, woody growth on the trunks or branches of trees that is sometimes used for making bowls, veneer, and coffee tables, such as from the redwood.

Burlap A loosely woven fabric made of jute or hemp used to protect newly seeded lawns from wind, water and birds.

Burn lime (CaO) A caustic solid used to neutralize acid soil or to raise the pH of the soil.

Burr knot A rough outgrowth, often present on the trunks or roots of certain trees.

Bursa Extension or flap of cuticle at the side of the male nematode sex organ, used for orienting during mating.

Bursiculate Resembling a small pouch, or having a small, pouch-like cavity.

Bush 1. A thicket; a clump of trees or shrubs.

2. A low and multi-branched shrub.

3. An expanse of forest or shrubby vegetation.

Bush-type Plant with determinate growth.

Butterfly garden A garden planted for the purpose of attracting butterflies by using plants whose flowers produce nectar for them. (i.e. butterfly bush -Buddleia and butterfly flower - Asclepias).

Buttonbuah Acommon name for Cephalanthus occidentalis, a North American shrub, on account of its globular flower heads.

C

C horizon Horizons or layers, excluding hard bedrock, that are little affected by pedogenic processes and lack properties of O, A, E, or B horizons.

C:N ratio The amount of carbon in organic material divided by the amount of total nitrogen in the same organic material.

Cactus This is the common name for plants in the group, Cactaceae. All cacti are succulent, although not all succulents are cacti. Cacti can be distinguished from other succulents by their areoles, or pad-like buds, from which shoots, flowers, and spines grow. Areoles aren't always easily seen.

Cadherins A group of cell adhesion molecules that form stable cell-cell junctions at adherens junctions and desmosomes.

Caenorhabditis elegans A nematode used as a simple multicellular model for development.

Caespitose Tufted; growing in small clumps.

Calcareous soil Soil containing carbonate of lime or limestone; chalky or limy.

Calcium An element needed by plants. It can be added to the soil.

Calcium carbonate A salt CaCo3 found in plant ashes, bones, shells, calcite and aragonite.

Calcium chloride A deliquescent salt CaC_{12} used as a drying and dehumidifying agent.

Calcium cyanamide A compound $CaCN_2$ used as a fertilizer and weed killer.

Calcium phosphate 1 A phosphate used as a fertilizer.

2. A naturally occurring phosphate of calcium occurring as the main constituent of phosphate rock.

Calcium sulfate A byproduct of manufacturing concentrated superphosphate; contains 15-18 percent sulfur.

Calcium-calmodulin A regulatory protein that has four high-affinity calcium binding sites. The calmodulin-calcium complex can bind to and activate several enzymes.

Calibration (soil test) The process of determining the crop nutrient requirement at different soil test values.

Caliche A term used to identify cemented deposits of calcium carbonate.

Callose Amorphous, hardened carbohydrate constituent of plant cell walls, commonly developing upon injury.

Callus 1. Any unusually hard, thickened area or swelling on a plant, such as the thickening that is formed over wounds (such as the base of a cutting) by which the inner tissues are protected and healing is effected.

2. A waxy or fleshy protuberance, as on a labellum (of an Orchid).

Calmodulin A calcium-binding protein.

Calvin cycle A series of reactions by which six molecules of CO_2 are converted into glucose.

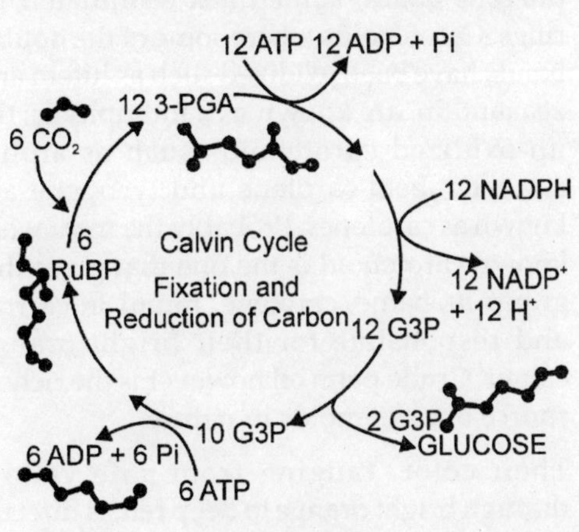

Fig. Calvin cycle

Calyptra A hood or covering on a flower or fruit. Also called calypter.

Calyptrate Furnished with a calyptra, as a capsule or a flower; resembling a calyptra, as a calyx that comes off like a lid.

Calyptrogen The cell layer from which the root cap originates.

Calyx The outer part of a flower, usually small and green but sometimes showy and brightly colored; it is formed from the sepals and encloses the petals in a bud.

Cambium A layer of tissue, one cell thick, formed between the wood and the bark of vascular plants that is capable of developing new cells. On one side, cambium develops into a layer of new wood and on the other side, into bark. At the same time, new cambium is produced. This process happens every year and increases the size of the trunk, as indicated by its annual rings when wood is cut into a cross section.

Camp phosphodiesterase An enzyme that degrades cyclic AMP.

Campanulate Bell-shaped.

Cane A long and slender, jointed, rigid, woody stem that is hollow or pithy, such as some grasses and palms and rattan, bamboo and sugarcane; the stem of raspberries or blackberries.

Canescent Growing white or gray; applied to plants with grayish-white, short, downy leaf hairs.

Canker A plant disease characterized (in woody plants) by the death of cambium tissue and loss and/or malformation of bark, or (in non-woody plants) by the formation of sharply delineated, dry, necrotic, localized lesions on the stem; "canker" may also be used to refer to the lesion itself, particularly in woody plants.

Canopy The expanded leafy top of a plant or plants.

CAP Common Agricultural Policy

Capillary 1. Water movement within the soil.

2. Water held by or found in the soil resulting from surface tension.

Capillary attraction Water moving against gravity and moving up to the soil's surface.

Capillary matting An interwoven fabric that wicks water from a reservoir. Used in propagators, seed flats, pots, or English baskets to provide plants with a constant supply of moisture.

Capillary porosity The volume of small pores in the soil that holds water.

Capitate Head-like; collected in a dense cluster.

Capitulum A dense head of flowers.

Capsid (coat protein) The protective layer of protein surrounding the nucleic acid core of a virus; the protein molecules which make up this layer.

Capsule A pod or seed vessel made of two or more cells, which becomes dry and splits open when mature to release its seeds.

Captan A man-made fungicide used to control many disease such as leaf spots and damping-off.

Carbaryl A man-made insecticide that kills insects (even beneficial ones) on contact. It is very poisonous to bees. Commonly called Sevin.

Carbohydrate A molecule with the formula $(CH_2O)n$. Carbohydrates include both simple sugars and polysaccharides.

Carbon One of the most common chemical elements found in the soil and air that is needed by living plants.

Carbon cycle The continuous circulation of carbon atoms from inorganic carbon dioxide (CO_2) to organic molecules and back to CO_2.

Carbon dioxide (CO_2) A colorless gas found in the air.

Carbon Dynamics Movement of carbon through an ecosystem.

Carbon/nitrogen (C/N) ratio The ratio of the mass of organic carbon to the mass of organic nitrogen in soil, organic material, plants, or microbial cells.

Carbonation Reaction with carbonic acid.

Carbonic acid A weak acid formed when water combines with carbon dioxide.

Carbonization Process of fossilization in which heat and pressure of the rock layers drive off all volatile elements from remains of organisms, leaving only carbon.

Carcinogen A substance or agent that causes cancer.

Carcinoma A cancer of epithelial cells.

Cardiolipin A phospholipid containing four hydrocarbon chains.

Carnivores Animals that feed on other animals or material of animal origin.

Carotenoid Carotenoids are organic pigments that are naturally occurring in plants and some other photosynthetic organisms like algae, some types of fungus and some bacteria. There are over 600 known carotenoids; they are split into two classes, xanthophylls and carotenes. Carotenoids are characterized by a large (35-40 carbon atoms) polyene chain, sometimes terminated by rings. Carotenoids where some of the double bonds have been oxidized such as lutein and zeaxanthin, are known as xanthophylls; the un-oxidized carotenoids such as alpha-carotene, beta-carotene and lycopene are known as carotenes. Probably the most well-known carotenoid is the one that gives this group its name, carotene, found in carrots and responsible for their bright orange colour. Crude palm oil however is the richest source of carotenoids in nature.

Their color, ranging from pale yellow through bright orange to deep red, is directly linked to their structure. The double carbon-carbon bonds interact with each other in a process called conjugation. As the number of double bonds increases, the wavelength of the absorbed light increases, giving the compound an increasingly red appearance.

In photosynthetic organisms, carotenoids play a vital role0auto-oxidation. In non-photosynthetic organisms, carotenoids have been linked to oxidation-preventing mechanisms.

Carotenoids have many physiological functions. Given their structure (above) carotenoids are efficient free-radical scavengers, and they enhance the vertebrate immune system. Consequently, epidemiological studies have shown that people with high beta-carotene intake and high plasma levels of beta-carotene have a significantly reduced risk of lung cancer. But studies of supplementation with large doses of beta-carotene in smokers have shown an increase in cancer risk (possibly because excessive beta-carotene results in breakdown products that reduce plasma Vitamin A and worsen the lung cell proliferation induced by smoke). Similar results have been found in other animals.

Animals are incapable of synthesizing carotenoids, and must obtain them through their diet, yet they are common and often in ornamental features. For example, the pink colour of flamingos and salmon, and the red colouring of lobsters are caused by carotenoids. Carotenoids are used in ornamental traits because, given their physioloigcal and chemical properties, they can be used as honest indicators of individual health, and hence they can be used by animals when selecting potential mates.

Products of carotenoid degradation such as ionones, damascones, and damascenones are also important fragrance chemicals that are used extensively in the perfumes and fragrance industry. Both beta-Damascenone and beta-Ionone although low in concentration in rose distillates are the key odor contributing compounds to flowers. In fact, the sweet floral smells present in black tea, aged tobacco, grape, and many fruits are due to the aromatics compounds resulting from carotenoid breakdown.

Carpel The female part of the flower. A carpel consists of three parts: 1. the ovary, a swollen basal portion containing the ovule, which becomes the seed after fertilisation,

2. styles, extensions of the ovary and

3. stigmas, specialised filaments on which the pollen falls and germinates.

Carpogenic Fruit-producing. Also carpogenous.

Carpogonium (carpogonia) In the Rhodophyta, the organ of the gametophyte that houses the egg; equivalent to oogonium.

Carpophore The wiry stalk that bears the carpels of some compound fruits.

Carposporophyte The first of two successive diploid generations in the Rhodophyta, developing from the zygote and producing carpospores.

Carrier proteins Proteins that selectively bind and transport small molecules across a membrane.

Carrying capacity The number of animals an area of land can support.

Caruncle An outgrowth surrounding the scar on a seed.

Caryopsis A small, one-seeded, dry fruit in which the fruit and seed are incorporated into a single grain, as in wheat and all other cereal grains.

Cash flow Movement of funds through a business

Casparian strip A ribbon-like layer of water-repellent suberin (waxy layer) that is wrapped around the tops, bottoms, and two side walls of endodermal cells in roots of plants

Caspases A family of proteases that bring about programmed cell death.

Casting Premature loss of abscised leaves or twigs.

Catabolic Degradative processes that generate energy for the cell through oxidation of organic molecules.

Catalase An enzyme capable of decomposing hydrogen peroxide into water and oxygen.

Catalyst A material that increases the rate of a chemical reaction without undergoing a chemical change.

Catalyzes Speeds up.

Cataphyll These are the rudimentary leaves that precede a stage of growth. (such as the cotyledons of an embryo, the scale-like leaf of a rhizome, bud, or at the base of a stem).

Catch crop A method of increasing garden productivity by filling in the empty spaces created when slower-growing vegetables are harvested with fast-growing crops. (i.e. green onions or radishes are grown in the space created when a broccoli plant is cut.

Catchments An area from which all the drainage water passes into one stream or other body of water.

Catena The soil that is formed from similar parent materials but doesn't have like soil characteristics because of the differences in drainage or relief.

Cation An ion that carries a positive charge of electricity like calcium, magnesium, sodium, potassium and hydrogen.

Cation exchange capacity (CEC) The ability of a soil to hold certain elements which have a positive charge; mainly a function of clay content and organic matter. Some plant nutrients, such as potassium and caldium are cations, so a soil with higher CEC will generally be more fertile because of its greater ability to hold these nutrients; measured in milliequivalents/100 grams.

Cation exchange The interchange between a cation in solution and another cation at the surface of a negatively charged material such as clay or organic matter.

Cationic Having positive charge.

Catkin A dense, scaly spike of flowers that fall after flowering or fruiting, as in the willow or birch; an ament.

Caudate Having a tail-like appendage; a "tail" or narrowed, apical extension of some sepals and petals.

Caudex The swollen water-storage tissue, usually consisting of both the root and stem of a succulent or woody plant.

Caudiciform Possessing a caudex; a plant with a swollen stem.

Caudicle Extensions of tissue derived from the anther and connected to pollinia in many Orchids.

Caulescent Having an obvious stem that grows above the soil's surface.

Caulocarpous Bearing a fruit repeatedly on the same stem; applied to plants that have perennial stems.

Causal agent Organism or agent that incites and governs disease or injury.

Caveolae Small invaginations of the plasma membrane that may be involved in endocytosis.

Cavitation Spontaneous occurrence of a vapor phase in a liquid under tension; occurs in xylem of plants that are under moderate to severe water-deficit stress

Cavities Tunnels left inside the corn stalk from European corn borer feeding

Cdc2 A protein-serine/threonine kinase that is a key regulator of mitosis in eukaryotic cells.

Cdk inhibitor (CKI) A family of proteins that bind Cdks and inhibit their activity.

Cdks Cyclin dependent protein kinases that control the cell cycle of eukaryotes.

cDNA (complementary DNA) DNA produced in vitro in such a way that the sequence of bases is complementary to a particular messenger RNA. Complementary DNA (cDNA) is used to study gene expression because it is more stable than RNA and is more amenable to recombinant cloning techniques.

cDNA library A collection of recombinant cDNA clones.

CEC-cation exchange capacity (CEC) The sum of exchangeable bases plus total soil acidity at a specific pH, values, usually 7.0 or 8.0. When acidity is expressed as salt extractable acidity, the cation exchange capacity is called the effective cation exchange capacity (ECEC) because this is considered to be the CEC of the exchanger at the native pH value. It is usually expressed in centimoles of charge per kilogram of exchanger (cmolckg-

1) or millimoles of charge per kilogram of exchanger.

Cell A small, usually microscopic mass of protoplasm, usually with one or more nuclei and some nonliving material, with an external membrane; the basic structural and functional unit of life.

Cell adhesion molecules Transmembrane proteins that mediate cell-cell interactions.

Cell cortex The actin network underlying the plasma membrane.

Cell cycle The cell cycle, or cell-division cycle (CDC), is the series of events in a eukaryotic cell between one cell division and the next. It consists of four phases: G_1 phase, S phase, G_2 phase (collectively known as interphase) and M phase. M phase is itself composed of two tightly coupled processes: mitosis, in which the cell's chromosomes are divided between the two daughter cells, and cytokinesis, in which the cell's cytoplasm physically divides. Cells that have temporarily or reversibly stopped dividing are said to have entered a state of quiescence called G_0 phase, while cells that have permanently stopped dividing due to age or accumulated DNA damage are said to be senescent.

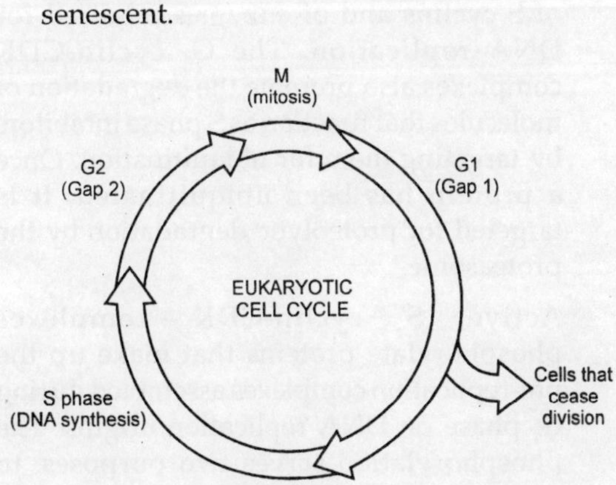

Fig. Cell Cycle

The molecular events that control the cell cycle are ordered and directional; that is, each process occurs in a sequential fashion and it is impossible to "reverse" the cycle. There are two key classes of regulatory molecules that determine a cell's progress through the cell cycle: cyclins and cyclin-dependent kinases.

Although the various stages of interphase are not usually morphologically distinguishable, each phase of the cell cycle has a distinct set of specialized biochemical processes that prepare the cell for entry into the next stage. It should be remembered that, throughout interphase, the cell carries out its normal metabolic activities and is actively engaging in transcription and translation of its genome. 1. In G_1 *phase*, the cell carries on its usual metabolic activities while preparing to duplicate its DNA. These preparations often include growing by increasing the amount of cytoplasm and the number of important organelles such as mitochondria. (This is particularly important in organisms and cell types that divide their cytoplasm unevenly, as in budding yeast.) In G_1 a diploid cell (such as a human cell) has a complement of *2N* chromosomes, where *N* is the gene copy number; in sexually reproducing organisms this amounts to one chromosome inherited from each parent. The actual quantity of DNA is described as 2c, where the "c" value is measured in picograms and 1c is equal to the quantity of DNA in a single haploid genome. The end of G_1 is demarcated by a "point of no return" beyond which the cell is committed to dividing; in yeast this is called START and in multicellular eukaryotes it is termed the restriction point.

2. In *S phase*, the cell duplicates its DNA.

3. In G_2 *phase*, the cell continues with growth and metabolism in preparation for undergoing mitosis. In this quantity of DNA within the cell has increased to *4c*, but the cell is still considered diploid.

4. In *M phase* the cell segregates its chromosomes so that both daughter cells receive a total complement of *2N*. The four

stages of mitosis - prophase, metaphase, anaphase, and telophase - also progress in a sequential and directional fashion, like the cell cycle as a whole. Telophase, the final stage of mitosis, is accompanied by cytokinesis; when the cytoplasm is completely divided, the cycle is complete and the new daughter cells are said to be in G_1 again. The exact mechanism of cytokinesis is highly organism- and cell type-dependent; for example, in plant cells surrounded by a rigid cell wall, cytokinesis occurs via the formation of a cell plate, while animal cells are "pinched" in two by a ring formed from a structural protein called actin.

Although the illustration assigns the four stages of the cell cycle roughly equal durations, a cell actually spends a very small amount of its time in M phase. The overall duration of the cell cycle depends on the organism and type of cell.

The term "post-mitotic" is sometimes used to refer to both quiescent and senescent cells. Nonproliferative cells in multicellular eukaryotes generally enter the quiescent G_0 state from G_1 and may remain quiescent for long periods of time, possibly indefinitely (as is often the case for neurons). This is very common for cells that are fully differentiated. Cellular senescence is a state that occurs in response to DNA damage or degradation that would make a cell's progeny nonviable; it is often a biochemical alternative to the self-destruction of such a damaged cell by apoptosis.

Cyclins and cyclin-dependent kinases (CDKs) are the two critical classes of molecules in regulation of cell cycle progression. Cyclins form the regulatory subunits and CDKs the catalytic subunits of an activated heterodimer; cyclins have no catalytic activity and CDKs are inactive in the absence of a partner cyclin. When activated by a bound cyclin, CDKs perform a common biochemical reaction called phosphorylation that activates or inactivates target proteins to orchestrate coordinated entry into the next phase of the cell cycle. Different cyclin-CDK combinations determine the downstream proteins targeted.

Many of the genes encoding cyclins and CDKs are conserved among all eukaryotes, but in general more complex organisms have more elaborate cell cycle control systems that incorporate more individual components. Many of the relevant genes were first identified studying yeast, especially *Saccharomyces cerevisiae*; genetic nomenclature in yeast dubs many of these genes *cdc* (for "cell division cycle") followed by an identifying number, e.g., *cdc28*. In the following discussion generic names such as "S cyclin" will be used to maintain generality, with the understanding that this may refer to one or to several homologous molecules in any given organism, and that some organisms may combine multiple functions in one molecule.

Upon receiving a pro-mitotic extracellular signal, G_1 cyclin-CDK complexes become active to prepare the cell for S phase, promoting the expression of transcription factors that in turn promote the expression of S cyclins and of enzymes required for DNA replication. The G_1 cyclin-CDK complexes also promote the degradation of molecules that function as S phase inhibitors by targeting them for ubiquitination. Once a protein has been ubiquitinated, it is targeted for proteolytic degradation by the proteasome.

Active S cyclin-CDK complexes phosphorylate proteins that make up the pre-replication complexes assembled during G_1 phase on DNA replication origins. The phosphorylation serves two purposes: to activate each already-assembled pre-replication complex, and to prevent new complexes from forming. This ensures that every portion of the cell's genome will be replicated once and only once. The reason for prevention of gaps in replication is fairly clear, because daughter cells that are missing all or part of crucial genes will die. However, for reasons related to gene copy number effects, possession of extra copies of certain genes would also prove deleterious to the daughter cells.

Mitotic cyclin-CDK complexes, which are synthesized but inactivated during S and G_2 phases, promote the initiation of mitosis by stimulating downstream proteins involved in chromosome condensation and mitotic spindle assembly. A critical complex activated during this process is a ubiquitin ligase known as the anaphase-promoting complex (APC), which promotes degradation of structural proteins associated with the chromosomal kinetochore. APC also targets the mitotic cyclins for degradation, ensuring that telophase and cytokinesis can proceed.

A molecular surveillance system monitors the cell's progress through the cell cycle and halts progression if extensive DNA damage has occurred or if a key event, such as the attachment of a chromosome to the mitotic spindle, has not occurred properly. These checkpoints help to ensure that a cell divides only when it has completed all of the molecular prerequisites for producing healthy daughter cells.

Three key stages of progression through the cell cycle involve the degradation of signaling molecules and are therefore irreversible; passage through these checkpoints not only "certifies" that the cell is capable of proceeding to the next step, but also commits the cell to that process. The restriction point that marks the transition from G_1 to S phase is the first such transition; the others occur between metaphase and anaphase (the spindle checkpoint) and between anaphase and telophase when mitotic cyclins are degraded.

The full system of checkpoints also includes monitoring of the cell's DNA for unrepaired damage and for successful completion of replication. 1. There are four DNA-damage checkpoints regulated primarily by the protein p53: one during G_1, one at the entry into S phase, one near the end of S phase, and one at the entry into M phase.

2. The unreplicated DNA checkpoint that occurs at the entry into M phase ensures that the cell has successfully and completely replicated its genome.

3. The spindle-assembly checkpoint occurs during anaphase and checks that all chromosomes have successfully attached to the mitotic spindle.

4. The chromosome-segregation checkpoint in telophase ensures that the chromosomes have properly migrated to the cell poles and that the cell is ready for cytokinesis and exit from mitosis.

The DNA-damage checkpoint protein p53 has been implicated in a large number of human cancers because its absence allows the cell to proceed into S phase with unrepaired DNA damage that leads to mutations when the DNA replication machinery misinterprets the damaged region or uses a lossy DNA repair mechanism to avoid the more serious consequence of proceeding through the cell cycle with a partially unreplicated genome. Very few of these mutations would by themselves be troublesome, but their continued accumulation in p53-deficient cell lineages produces a high likelihood of introducing mutations in oncogenes, many of which are associated with controlling cell division and preventing unconstrained growth. p53 also serves as a mechanism to induce apoptosis, or "cell suicide", in cases where the cell has sustained irreparably high amounts of DNA damage.

Cell membrane The thin layer of protein and fat that surrounds the cell, but is inside the cell wall. The cell membrane is semipermeable, allowing some substances to pass into the cell and blocking others.

Cell membrane disrupters Mode of action of a group of herbicides that kill by destroying cell integrity.

Cell wall A cell wall is a more or less solid layer surrounding a cell. They are found in bacteria, archaea, fungi, plants, and algae. Animals and most other protists have cell membranes without surrounding cell walls. When a cell wall is removed using cell wall degrading enzymes, what is left of the cell and its surrounding plasma membrane is called a protoplast. The cell wall's main purpose is to actually protect the interior from any physical movement that may damage the cell.

Plant cell walls have a number of functions: they provide rigidity to the cell for structural and mechanical support, maintaining cell shape, the direction of cell growth and ultimately the architecture of the plant. The cell wall also prevents expansion when water enters the cell. The term turgor is used to describe this pressure that is induced by excess water inside the plant cell. Cell walls protect against pathogens and the environment and are a store of carbohydrates for the plant. The cell wall is constructed primarily from a carbohydrate polymer called cellulose.

The middle lamella is laid first, formed from the cell plate during cytokenesis, and the primary cell wall is then expanded inside the middle lamella. The actual structure of the cell wall is not clearly defined and several models exist - The covalently linked cross model, the tether model, the diffuse layer model and the stratified layer model are to name a few. However, the primary cell wall, can be defined as composed of cellulose microfibrils aligned at all angles. Microfibrils are held together by hydrogen bonds to provide a high tensile strength. The cells are held together and share the gelatinous membrane called the *middle lamella*, which contains magnesium and calcium pectates (salts of pectic acid). Cells interact though plasmodesma(ta), which are inter-connecting channels of cytoplasm that connect to the protoplasts of adjacent cells across the cell wall.

In some plants and cell types, after a maximum size or point in development has been reached, a *secondary wall* is constructed between the plant cell and primary wall. Unlike the primary wall, the microfibrils are aligned mostly in the same direction, and with each additional layer the orientation changes slightly. Cells with secondary cell walls are rigid. Cell to cell communication is possible through *pits* in the secondary cell wall that allow plasmodesma to connect cells through the secondary cell walls.

The major carbohydrates making up the primary cell wall are cellulose, hemicellulose and pectin. The cellulose microfibrils are linked via hemicellulosic tethers to form the cellulose-hemicellulose network, which is embedded in the pectin matrix.

Plant cells walls also incorporate a number of proteins; the most abundant include hydroxyproline-rich glycoproteins (HRGP), also called the extensins, the arabinogalactan proteins (AGP), the glycine-rich proteins (GRPs), and the proline-rich proteins (PRPs). With the exception of glycine-rich proteins, all the previously mentioned proteins are glycosylated and contain hydroxyproline (Hyp). Each class of glycoprotein is defined by a characteristic, highly repetitive protein sequence. Chimeric proteins contain two or more different domains, each with a sequence from a different class of glycoprotein. Most cell wall proteins are cross-linked to the cell wall and may have structural functions.

Secondary cell walls may contain lignin and suberin, making the walls rigid. The relative composition of carbohydrates, secondary compounds and protein varies between plants and between the cell type and age.

Cellular endosperm The endosperm formed at the base of the embryo sac from divisions of the fused polar nucleus. This region of endosperm nurtures the embryo during its first divisions.

Cellular path Water moves from cell to cell using membranes, either across membranes or using plasmodesmata

Cellulose A polysaccharide of glucose units that constitutes the chief part of the cell walls of plants .

Center of origin Geographical area where a plant originated.

Central dogma The concept that genetic information flows from DNA to RNA to proteins.

Centriole The intracellular organelle that represents an inactivated basal body. Centrioles do not occur in the Kingdom Fungi and the divisions Coniferophyta, Gnetophyta, and Anthophyta of the Kingdom Plantae.

Centromere A specialized chromosomal region that connects sister chromatids and attaches them to the mitotic spindle.

Centrosome (also called the "microtubule organizing center") A small body located near the nucleus it has a dense center and radiating tubules. The centrosomes is where microtubules are made. During cell division (mitosis), the centrosome divides and the two parts move to opposite sides of the dividing cell.

Cephalium A densely spined crown, which produces flowers and fruit. It is produced only by species Discocactus and Melocactus.

Ceraceous Waxy; applied to bodies that have the texture and color of new wax, as the pollen masses of many orchids.

Cernuous Drooping; pendulous; having the top curved or bent down.

Certification Seeds, propagative plant material, or nursery stock produced and sold under inspection to maintain genetic identity and purity, freedom from harmful pathogens, insect pests, and weed seeds. It is approved and certified by an official certifying agency.

Certified organic foods Organic" is a labeling term that denotes products produced under the authority of the Organic Foods Production Act. The term "organic" details the methods, practices and substances that can be used in producing and handling organic crops and livestock, as well as processed products. Organic agriculture practices cannot ensure that products are completely free of residues, however methods are used to minimize pollution from air, soil, and water.

CFU (abbr. for colony forming unit) The number of colonies formed per unit of volume or weight of a cell or spore suspension.

cGMP phosphodiesterase An enzyme that degrades cGMP.

Chaff The glumes, husks, or other pieces of the leaves and stems of wheat, oats, or other grains and grasses, especially when separated from the seed by threshing and winnowing.

Chalaza The basal region of the ovule where the stalk is attached.

Channel proteins Proteins that form pores through a membrane.

Channeled Hollowed out; trough-like; applied to petioles, leaves, etc.

Channels A transport protein in the membrane of cells that are selective pores, through which molecules can diffuse across the membrane.

Chaperone A protein that facilitates the correct folding or assembly of other proteins.

Chaperonin A family of heat-shock proteins within which protein folding takes place.

Chelates A chemical compound that combines a metallic atom with a molecule using multiple chemical bonds.

Chemical drenches Can be used to control soil-borne disease.

Chemical gardening A method of cultivating plants in which chemicals are used to control undesirable pests or weeds.

Chemiosmotic coupling The generation of ATP from energy stored in a proton gradient across a membrane.

Chemosynthesis The process by which bacteria use energy from chemicals, such as hydrogen sulfide, to combine water and carbon dioxide to produce carbohydrates

Chemotaxis (Chemotropism) Movement or growth of an organism in response to changing concentration of a chemical stimulus, often in relation to food or for mating.

Chemotherapy Treatment of plant disease with chemicals (e.g. antibiotics or fungicides) absorbed and translocated internally.

Chert A structureless form of silica.

Chestnut soil Dark brown surface soil.

Chiasmata Sites of recombination that link homologous chromosomes during meiosis.

Chicken manure Strong organic fertilizer derived from chicken droppings. Also called hen manure.

Chimera A plant consisting of two or more genetically distinct tissues or an artificially produced plant having tissues of several species.

Chimeric plant A transformed plant that has the new gene in only some of its cells.

Chipper A tool made of a chute, blades and an electric or gas motor; used for turning branches into wood chips.

Chisel A type of minimum tillage that breaks up the aggregates of the soil more than discing.

Chitin A natural pesticide made from pulverized crustacean shells. This is added to the soil to increase the feeding of soil organisms on nematode eggs, which helps control these pests.

Chlamydospore Thick-walled or double-walled asexual resting spore formed from hyphal cells (terminal or intercalary) or by transformation of conidial cells that can function as an overwintering stage.

Chloroacetamides Family of herbicides that inhibit shoot inhibition and belong to the seedling growth inhibitor mode of action.

Chlorophyll Chlorophyll is a green photosynthetic pigment found in most plants, algae, and cyanobacteria. Its name is derived from ancient Greek: *chloros* = green and *phyllon* = leaf. Chlorophyll absorbs most strongly in the blue and red but poorly in the green portions of the electromagnetic spectrum, hence the green color of chlorophyll-containing tissues like plant leaves.

Chlorophyll is an essential component of photosynthesis, which helps plants get energy from the light. Chlorophyll molecules are specifically arranged in and around pigment protein complexes called photosystems, which are embedded in the thylakoid membranes of chloroplasts. In these complexes, chlorophyll serves two primary functions. The function of the vast majority of chlorophyll (up to several hundered per photosystem) is to absorb light and transfer that light energy by resonance energy transfer to a specific chlorophyll pair in the reaction center of the photosystems. Photosystem II and Photosystem I have their own distinct reaction center chlorophylls, named P680 and P700, respectively. These pigments are named after the wavelength (in nanometers) of their red-peak absorption maximum. The identity, function and spectral properties of the chlorophylls in each photosystem are distinct and determined by each other and the protein structure surrounding them. Once extracted from the protein into a solvent (such as acetone or methanol), these chlorophylls lose those distinctions and become a homogenous mixture of identical molecules.

The function of the reaction center chlorophyll is to use the light energy absorbed by and transferred to it from the other chlorophylls in the photosystems to undergo a charge separation, a specific redox reaction in which the chlorophyll donates an electron into a series of molecular intermediates called an electron transport chain. The charged reaction center chlorophyll ($P680^+$) is then reduced back to its ground state by accepting an electron. In Photosystem II, the electron which reduces $P680^+$ ultimately comes from the oxidation of water into O_2 and H^+ through several intermediates. This reaction is how photosynthetic organisms like plants produce O_2 gas, and is the source for practically all the O_2 in Earth's atmosphere. Photosystem I typically works in series with Photosystem II, thus the $P700^+$ of Photosystem I is usually reduced, via many intermediates in the thylakoid membrane, by electrons ultimately from Photosystem II. Electron transfer reactions in the thylakoid membranes are complex, however, and the source of electrons used to reduce $P700^+$ can vary. The electron flow produced by the reaction center chlorophylls is used to shuttle H^+ ions across the thylakoid membrane, setting up a chemiosmotic potential mainly used to produce ATP chemical energy, and those electrons ultimately reduce NAD^+ to NADPH a universal reductant used to reduce CO_2 into sugars as well as for other biosynthetic reductions.

Reaction center chlorophyll-protein complexes are capable of directly absorbing light and performing charge separation events without other chlorophylls, but the absorption cross section (the likelihood of absorbing a photon under a given light intensity) is small. Thus, the remaining chlorophylls in the photosystem and antenna pigment protein complexes associated with the photosystems all cooperatively absorb and funnel light energy to the reaction center. Besides chlorophyll *a*, there are other pigments, called accessory pigments, which occur in these pigment-protein antenna complexes. They include other forms of chlorophyll, such as chlorophyll *b* in green algal and higher plant antennae, while other algae may contain chlorophyll *c* or *d*. In addition, there are many non-chlorophyll accessory pigments, such as carotenoids or phycobilins which also absorb light and transfer that light energy to the photosystem chlorophylls. Some of these accessory pigments, particularly the carotenoids, also serve to absorb and dissipate excess light energy, or work as antioxidants. The large, physically associated group of chlorophylls and other accessory pigments is sometimes referred to as a *pigment bed*, though this term is losing prominence with the advent of detailed knowledge of the structural organization of the photosystem and antenna complexes.

The different chlorophyll and non-chlorophyll pigments associated with the photosystems all have different spectra, either because the spectra of the different chlorophylls are modified by their local protein environment, or because the accessory pigments have intrinsically different absorption spectra from chlorophyll. The net result is that, *in vivo* the total absorption spectrum is broadened and flattened such that a wider range of red, orange, yellow and blue light can be absorbed by plants and algae. Most photosynthetic organisms do not have pigments which absorb green light well, thus most remaining light under leaf canopies in forests or under water with abundant plankton is green, a spectral effect called the "green window". Some organisms, however, contain phycobilin pigments of cyanobacteria and red algae can absorb green light relatively well and they can exploit the little remaining green light in these habitats.

Chlorophyll is a chlorin pigment, which is structurally similar to and produced through the same metabolic pathway as other porphyrin pigments such as heme. At the center of the porphyrin ring is a magnesium ion. The chlorin ring can have several different side chains, usually including a long phytyl chain. Historical evidence for the importance of chlorophyll in photosynthesis

A simple experiment can show how chlorophyll is associated with photosynthesis. After destarching a leaf from a variegated plant and exposing it to light for several hours, starch can be seen to rapidly accumulate again by staining with iodine solution. Variegated leaves have green areas that contain chlorophyll and white areas that do not. The iodine-stained starch only accumulates in regions of the leaf that were green and therefore contained chlorophyll. This shows that photosynthesis does not occur in areas where chlorophyll is absent, and constitutes evidence that the presence of chlorophyll is a requirement for photosynthesis.

Chloroplast Chloroplasts are organelles found in plant cells and eukaryotic algae that conduct photosynthesis. Chloroplasts capture light energy from the sun to produce the free energy stored in ATP and NADPH through a process called photosynthesis.

Chloroplasts are one of the forms a plastid may take, and are generally considered to have originated as endosymbiotic cyanobacteria. In this respect they are similar to mitochondria, but are found only in plants and protista. Both organelles are surrounded by a double celled composite membrane with an intermembrane space; both have their own DNA and are involved in energy metabolism; and both have reticulations, or many infoldings, filling their inner spaces.

In green plants chloroplasts are surrounded by two lipid-bilayer membranes. The inner membrane is now thought to correspond to the outer membrane of the ancestral cyanobacterium. The chloroplast genome is considerably reduced compared to that of free-living cyanobacteria, but the parts that are still present show clear similarities. Many of the missing genes are encoded in the nuclear genome of the host.

It is interesting to note that in some algae (such as the heterokonts and other protists such as Euglenozoa and Cercozoa), chloroplasts seem to have arisen through a secondary event of endosymbiosis, in which a eukaryotic cell engulfed a second eukaryotic cell containing chloroplasts, forming chloroplasts with three or four membrane layers. In some cases, such secondary endosymbionts have themselves been engulfed by still other eukaryotes, forming tertiary endosymbionts.

Chloroplasts are flat discs usually 2-10 micrometer in diameter and 1 micrometer thick. The chloroplast has a two membrane envelope termed the Inner & Outer membrane respectively. Between these two layers is the Intermembrane space.

The fluid within the chloroplast is called the stroma, corresponding to the cytoplasm of the bacterium, and contains tiny circular DNA and ribosomes, though most of their proteins are encoded by genes contained in the cell nucleus, with the protein products trafficked to the chloroplast.

Within the stroma are stacks of thylakoids, the sub-organelles where photosynthesis actually takes place. A stack of thylakoids is called a granum (plural: grana). A thylakoid looks like a flattened disk, and inside is an empty area called the thylakoid space or lumen. The photosynthesis reaction takes place on the membrane of the thylakoid, and, as is also the case with mitochondria, involves the coupling of cross-membrane fluxes with biosynthesis.

Chlorosis Failure of chlorophyll development, caused by disease or a nutritional disturbance; fading of green plant color to light green, yellow, or white.

Chlorothalonil A multi-purpose fungicide used to treat a variety of plant diseases. Commonly known as Daconil.

Cholesterol A lipid consisting of four hydrocarbon rings. Cholesterol is a major constituent of animal cell plasma membranes and the precursor of steroid hormones.

Choripetalous Having the petals unconnected.

Chorisis The multiplication, by congenital division during development, of a leaf or floral organ that is normally whole.

Christae (singular crista) The multiply-folded inner membrane of a cell's mitochondrion that are finger-like projections. The walls of the cristae are the site of the cell's energy production (it is where ATP is generated).

Chromatid One of the two replicated strands of DNA and associated proteins forming a chromosome following replication.

Chromatin The fibrous complex of eukaryotic DNA and histone proteins. See histones, nucleosome, and chromatosome.

Chromatosome A chromatin subunit consisting of 166 base pairs of DNA wrapped around a histone core and held in place by a linker histone.

Chromosome A unit of hereditary material composed of 1000's of genes. During cell division chromosomes duplicate and divide, maintaining the genetic integrity of the organism.

Chromosome map A genetic map showing the location of genes on a chromosome.

Chromosomes Thread-shaped structures that occur in pairs in the nucleus of a cell. Chromosomes contain the genetic information of living organisms.

Chronic Of long duration; continuing. Lasting for a long period of time or marked by frequent recurrence.

Chronic toxicity Poisoning due to low levels of exposure to a compound over a period of time .

Chronosequence A sequence of related variables (e.g., forest stands, soils) that differ from one another in certain properties primarily as a result of time. A set of stands of trees, in which one was recently clearcut, another had been cut 5 years ago, another 15 years ago, another 25 years ago is an example of a chronosequence.

Chytridiomycetes A fungus group, also called the Chytridiomycota, characterized by the production of motile zoospores and resting sporangia; the plant-pathogenic species are all biotrophs that are restricted to the cells of their host.

Ciliate Marginally fringed with hairs.

Ciliolate Marginally fringed with tiny hairs.

Cilium A microtubule-based projection of the plasma membrane that moves a cell through fluid or fluid over a cell.

Cincinnus A form of definite flowering in which the successive axes arise alternately to the right and left of the preceding one, in distinction from the bostryx, in which the growth is all on one side; a scorpioid cyme.

Circinate Circular or ring-shaped. Applied to leaves that are rolled up on itself from the tip toward the bottom, like a shepherd's cane. (i.e. fern fronds and the leaves of the Sundew -Drosera)

Circinnate vernation Unfolding of a newly-formed leaf in a spiral structure.

Circulative transmission (persistent transmission) Virus transmission characterized by a long period of acquisition of the virus by a vector (typically an insect), a latent period of several hours before the vector is able to transmit the virus, and retention of the virus by the vector for a long period, usually several days; the virus circulates in the body of the vector.

Circulative-propagative transmission (propagative transmission) Pathogen transmission characterized by a long period of acquisition of the pathogen (usually a mollicute, e.g. phytoplasma or spiroplasma, and sometimes a virus) by a vector (typically an insect), a latent period before the vector is able to transmit the pathogen, and retention of the pathogen by the vector for a long period because the pathogen reproduces or replicates in the vector.

Circumnutation The continuous motion of some part of a plant, as the top of the stem or a tendril, in which it describes irregular ellipses or circles.

Circumscissile Opening or divided by a horizontal circular line; applied to pods or seed capsules that open with a lid, like a box. The fruit in such cases is called pyxidium.

Cirrhus A curled, tendril-like mass of exuded spores, held together by a slimy matrix.

Cirrose 1. A leaf tipped with a tendril, or, in mosses, with a very narrow or hair-like sinuous point.

2. Resembling tendrils or coiling like them. Also cirrate, cirrous, cirrhous, cirrhose.

Cirrus A tendril; a long thread-like part by which some plants climb. Also cirrhus.

Cis acting control element A regulatory DNA sequence that serves as a protein binding site and controls the transcription of adjacent genes.

Citric acid cycle A series of reactions in which acetyl CoA is oxidized to CO

2. The central pathway of oxidative metabolism.

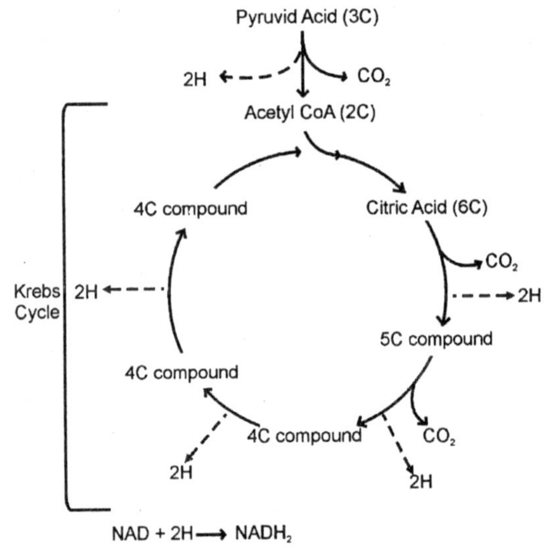

Fig. Citric acid cycle

CLA Country Landowners Association

Cladistic classification Classification of organisms based on evolutionary kinship.

Cladocarpous Having the fruit at the end of short side branches.

Cladode Flattened leaf-like stem.

Cladogram A graphic hypothesis of the phylogeny of a group of organisms, based on the distribution of shared homologies in the group.

Cladoptosis The annual falling of leafy twigs instead of individual leaves. (An example of this process takes place in many trees of the Cypress family.)

Clamp connection A bridge- or buckle-hyphal protrusion in basidiomycetous fungi, formed at cell division and connecting the newly divided cells.

Clasping Wholly or partially surrounding the stem.

Classification Arrangement of entities into orderly groups.

Classification of plants Taxonomy; the systematic naming of plants devised by Carl von Linne.

Clathrin A protein that coats the cytoplasmic surface of cell membranes and assembles into basketlike lattices that drive vesicle budding.

Clavate Club-shaped. Growing gradually thicker toward the far end. Also clavated.

Claw 1. The long and narrow base of a sepal or petal (including labellum), as in pinks, wallflowers, and Orchids.

2. A sharp tool usually having three tines, used to weed and cultivate.

Clay 1. A type of soil.

2. Soil that contains 10% or more clay.

3. A soil particle less than 0.002 mm in diameter.

Clay loam Soil containing 27% to 40% clay.

Clay mineral A naturally occurring inorganic crystalline material with particles less than 0.002 mm in diameter.

Clay pan 1. Soil rich in clay.

2. Soil hard when dry, stiff when wet and it will separate abruptly from overlying soil.

Clayey soil Soil made up of fine particles that won't hold much air.

Clearcut To remove or cut all trees in a tract of timber at one time.

Cleistogamy When a plant has regular, fully developed flowers as well as flowers in which development is checked in the bud, but are still fertile and produce a lot of seed. These undeveloped flowers are inconspicuous, non-opening, and lack petals, scent or nectar. They have small anthers that contain a few pollen grains and a dwarfed pistil. They are always fertile while the more perfect flowers of the same plant are usually nearly or completely sterile. Cleistogamy is known to happen in about 60 groups belonging to many different families, including Dicotyledonous. The violet is a good example .

Cleistothecium A spherical ascocarp that is closed at maturity.

Climate The accumulation of daily and seasonal weather events over a long period of time. The sum of all statistical weather information that helps describe a place or region.

Climber A plant that clambers upward by attaching itself to other plants or objects (such as a trellis). Climbing plants are distinguished as stem climbers, which wind upward around an erect support and, as tendril climbers, which cling to nearby objects by slender, coiling tendrils. Other plants climb and cling by means of spines or bristles, or by rootlets.

Clinandrium A cavity, at the apex of a column in orchids, in which the anthers rest.

Cloche A bell-shaped cover used to protect a plant from frost or cold; an individual cold frame.

Clone A population of cells or organisms of identical genotype; population of recombinant DNA molecules all carrying the same inserted sequence;

The vegetative propagation of an organism to produce a population of identical individuals; the use of in vitro recombination techniques to insert a particular DNA sequence into a vector.

Cloned gene A copy of a gene that has been combined with a plasmid and placed into a bacteria cell.

Cloning Producing a stock of individuals all derived asexually from one sexually produced

Coal ball A carbonate rock surrounding carbonized plant (or rarely animal) material, generally found as an inclusion in coal strata.

Coalesce Grow together into one body or spot.

Coalescent Growing together; uniting so as to form one body. Applied to the organic cohesion of similar parts.

Coarse sand A soil particle that measures between .50 to 1.0 millimeters.

Coat protein (capsid) The protective layer of protein surrounding the nucleic acid core of a virus; the protein molecules which make up this layer.

Coccus One of the separate divisions of a divided seedpod, which splits up into one-seeded cells.

Cochlear Shell-shaped.

Coding region Region of the DNA sequence between the promoter and the termination sequence. It contains the instructions about how to make a specific protein.

Codominance The situation in which both alleles in a heterozygous organism are equally expressed resulting in a phenotype having characteristics from both alleles.

Codon Three consecutive nucleotides in an RNA molecule that code for a single amino acid.

Codon bias The frequency with which individual codons are used by an organism.

Coelomycetes A group of fungi in the deuteromycetes (Imperfect Fungi) that produce pycnidia or acervuli .

Coenocytic Condition in which an organism consists of filamentous cells with large central vacuoles, and whose nuclei are not partitioned into separate compartments. The result is a long tube containing many nuclei, with all the cytoplasm at the periphery.

Coenocytic endosperm A multinucleate condition in which nuclei divide within a common cytoplasm without being separated from each other by a plasmalemma or cell wall.

Coenocytic Having multiple nuclei embedded in cytoplasm without cross walls; nonseptate

Coenzyme A Coenzyme A (CoA, CoASH, or HSCoA) is a coenzyme, notable for its role in the synthesis and oxidization of fatty acids, and the oxidation of pyruvate in the citric acid cycle. It is adapted from β-mercaptoethylamine, panthothenate and adenosine triphosphate.

The main function of coenzyme A is to carry acyl groups (such as the acetyl group) or thioesters. A molecule of coenzyme A carrying an acetyl group is also referred to as acetyl-CoA. It is sometimes referred to as 'CoASH' or 'HSCoA' because when it is not attached to a molecule such as an acetyl group, it is attached to a thiol group, -SH.

Acetyl-CoA is an important molecule itself. It is the precursor to HMG CoA, which is a vital component in cholesterol and ketone synthesis. Furthermore, it contributes an acetyl group to choline to produce acetylcholine, in a reaction catalysed by choline acetyltransferase. Its main task is conveying the carbon atoms within the acetyl group to the citric acid cycle to be oxidized for energy production.

The conversion of pyruvate into acetyl-CoA is referred to as the Pyruvate Dehydrogenase Reaction. It is catalyzed by an enzyme-complex called pyruvate dehydrogenase. The enzyme consists of 60 subunits: 24 pyruvate dehydrogenase, 24 dihydrolipoyl transacetylase, and 12 dihydrolipoyl dehydrogenase (commonly denoted E1, E2, and E3). 24 pyruvate dehydrogenase has the coenzyme TPP (thiamin pyrophosphate) incorporated into it, 24 dihydrolipoyl transacetylase has lipoate and coenzyme A, and 12 dihydrolipoyl dehydrogenase has the coenzymes FAD and NAD^+. Through a complex reaction, pyruvate is decarboxylated and turned into acetaldehyde, then attached to coenzyme A while NAD^+ is subsequently reduced to NADH and H^+.

Coenzyme A is very central to the balance between carbohydrate metabolism and fat metabolism. Normally, acetyl-CoA from fatty acid metabolism feeds into the citric acid cycle, contributing to the cell's energy supply. In the liver, when levels of circulating fatty acids are high, the production of acetyl-CoA from fat breakdown exceeds the cellular energy requirements. To make use of the energy available from the excess acetyl-CoA, ketone bodies are produced which can then circulate in the blood.

In some circumstances this can lead to an excess of ketone bodies in the blood, a condition known as ketosis. This can occur in diabetes, starvation or in people following low-carbohydrate diets, all of which can cause fats to be metabolised as a major source of energy.

Coenzyme Q A small lipid-soluble molecule that carries electrons between protein complexes in the mitochondrial electron transport chain.

Coenzymes Low-molecular-weight organic molecules that work together with enzymes to catalyze biological reactions.

Cofactor The nonprotein component of an enzyme. They often carry out activities for which amino acids are not suited.

Cohesins A complex of proteins that maintain the connection between sister chromatids.

Cohesion A molecular attraction of molecules

Cohesive Able to stick together tightly

Colcemid A drug that inhibits the polymerization of microtubules.

Colchicine A drug that inhibits the polymerization of microtubules.

Cold frame A bottomless box consisting of a wooden or metal frame with a glass or polyethylene top. It is placed on the ground over plants to protect them from cold or frost.

Coleoptile A protective cylindrical sheath-like structure that surrounds the shoot apex in cereal and grass embryos.

Coleorhiza The protective layers of tissue surrounding the primordial root (radicle) within the grain.

Collagen The major structural protein of the extracellular matrix.

Collar The portion of the seedling or plant near the surface of the soil; in grafted woody plants, the scion portion of the plant near the soil surface.

Collenchyma A layer of tissue right beneath the epidermis, having the cells thickened at the angles with a pad-like mass, which swells in water. It is found mainly in the young stems, petioles and leaf veins of many dicotyledonous plants.

Colloid soil Organic or inorganic tiny crystals that are small enough to be easily carried by air or water.

Colluvium Soil that is deposited at the base of a steep slope.

Colonial Condition in which many unicellular organisms live together in a somewhat coordinated group. Unlike true multicellular organisms, the individual cells retain their separate identities, and usually, their own membranes and cell walls.

Colonization Establishment and ramification of a pathogen within a host plant.

Colonize To infect and ramify through plant tissue with the growth of a pathogen.

Colony A group of identical cells (or bacteria) that develop from a single cell when the cell divides multiple times. The cells are considered 'clones' of the original cell.

Colony forming unit (abbr. cfu) The number of colonies formed per unit of volume or weight of a cell or spore suspension.

Columella (columellae) a central cylinder or dome of sterile tissue in the center of a sporangium.

Column An organ formed by the union of filaments with one another, as in Malvacea, or of stamens with the style and stigma, as in Orchids.

Columnar Tall and cylindrical or tapering.

Column-foot A basal extension of the column to which the labellum is attached, often flexibly.

Column-wings Paired protuberances or wing-like flanges on the column.

Coma The leafy head of a tree (such as Palms) or a tuft of leaves or bracts terminating a stem (such as the leafy top of a Pineapple). The silky hairs at the end of some seeds (such as fireweed - Epilobium).

Commissure A joint, seam, suture, or closure; the place where two bodies or parts of a body meet or unite. The face by which one carpel or mericarp attaches to another, as in Umbelliferae.

Common Name The Common Name describes the chemical name of the active ingredient of the herbicide.

Community (plant community) An assemblage of plants occurring together at any point in time, while denoting no particular ecological status. A unit of vegetation.

Compact Pressing air out of the soil.

Compaction The process by which soil particles are rearranged to decrease void space and bring them into closer contact with one another, thereby increasing the bulk density.

Companion cell Specialized cell adjacent to sieve tubes in phloem.

Companion planting/cropping A system of arranging plants in a way that a given species benefits from some characteristic of another. For example, it is thought that Tomatoes are protected from certain insects by the odor of Marigolds.

Compartmentalization Isolation of a specific tissue area by host barrier tissues.

Compartmentalize To separate with a barrier. This can mean a physical barrier such as a membrane, or two immiscible solvents such as oil and water

Competition Use or defense of a resource by one individual that reduces the availability of that resource to other individuals.

Competitive inhibitor A molecule which has a similar structure as the normal substrate an enzyme acts upon. A competitive inhibitor inhibits an enzyme's activity by directly competing with the normal substrate in binding to the active site on the enzyme.

Complanate Lying in one plane. Applied to leaves, especially of Mosses.

Complementary Term used in science to refer to the normal nucleotide-pairing in DNA replication, with cytosine pairing with guanine and adenine pairing with thymine.

Complete flower A flower with sepals, petals, stamens, and pistils.

Complete resistance A resistant reaction in which some aspect of disease development, usually symptom expression or pathogen reproduction, is completely stopped .

Complex soil 1. A pattern of soil.

2. A mixture of different kinds of soil whose areas are too small to be shown individually.

Composite Member of the Asteraceae family

Compost A mixture of organic residues and soil that is allowed to decompose biologically.

Compost activator A substance, usually bacterial, that is used to speed the breakdown of organic waste into compost. Also called activator or compost inoculent.

Compost aerator An elongated, flanged T-shaped device used to create passages for air and moisture in a compost pile. Also created compost turner.

Compost bin A container to hold decomposing garden and yard wastes, fertilizer, peat, lime, and other substances during the composting process.

Compost inoculant A substance that may contain bacteria, fungi, enzymes, and hormones that is used to speed the composting process and to improve the quality of the compost. Also called activator or compost activator.

Compost tea A liquid fertilizer made of compost dissolved in water.

Compost thermometer A thermometer having a long probe used to measure the temperature in the center of the compost pile.

Composter A container utilizing air, moisture, and solar heat to turn grass clippings, garden waste, and kitchen scraps into compost. Composters may consist of stacked trays or other chambers arranged to facilitate periodic turning of the mixture.

Compound Made up of two or more similar parts (i.e. a compound leaf).

Compound leaf A leaf that is divided into two or more leaflets.

Compressed Flattened.

Compressibility The susceptibility of a soil to decrease in volume when subjected to load (pressure).

Compression fossil The organic remains of an organism which have been pressed flat between rock strata.

Concentrate superphosphate Manufactured by reacting phosphoric acid with finely ground rock phosphate; often used in bulk blended products.

Concentric One circle within another with a common center.

Conceptacle a chamber at the tip of a branch of a rockweed in the Phaeophyta that contains the oogonia and antheridia.

Concrescence The union of cell walls by means of a cementing substance formed in process of growth so that they are inseparably grown together. Also called cementation.

Concrete A compact mass of sand, gravel, coarse pebbles, or stone chips cemented together to form a smooth surface for steps, patios, or walkways.

Concretions Pellet-like grains from concentrations of compounds in soils that cement the soil grains together.

Conditional Lethal Genes Genes that are expressed in response to a specific environmental stimulus such as a specific chemical. When expressed, it is lethal to the plant.

Conditioner A material added to a fertilizer that keeps it flowing free.

Conductance The power to conduct.

Conductivity The physical quantity that measures the readiness with which a medium transmits electricity. Expressed in micromhos per centimeter at 25° C.

Conduit A channel through which something (as a fluid) is conveyed.

Conduplicate Folded once longitudinally down the middle.

Cone 1. A flower or cluster of flowers having a cone-like shape.

2. A dry, usually elongated, multiple fruit formed of compactly overlapped scales (such as those of Pines, Firs and Spruces), in which a pair of naked seeds is borne upon the upper side of each scale (technically called a strobilus or strobile).

Conglomerate Densely clustered.

Conical Cone-shaped.

Conidiogenesis The formation of asexual spores (conidia).

Conidiogenous Producing and bearing conidia.

Conidioma Specialized conidia-bearing structure, e.g. acervulus, pycnidium, sporodochium, synnema.

Conidiophore Simple or branched hypha on which conidia are produced .

Conidiospore See conidium.

Conidium An asexual, nonmotile fungal spore that develops externally or is liberated from the cell that formed it.

Conifer A plant that produces cones such as a plant belonging to the family Coniferae, such as Pines, Cypresses, Junipers and Cedars.

Conjugated double bonds Double bonds on alternating pairs of carbon molecules in a linear or cyclic organic molecule. An example is as follows -C=C-C=C-C=C- The electrons in conjugated double bonds form a resonance structure that is smeared over the double bond system. These electrons are responsible for light absorption.

Conjugation Attachment of a molecule (herbicide) to another molecule like sugar, lipid, lingnin, cellulose, which inactivates the original molecule.

Conk A shelf-like, typically hardened basidiocarp of a wood decaying fungus, usually a polypore.

Connate United congenitally.

Connation Fusion of like parts; e.g. sepal with sepal; contrasted with adnation.

Conservation Protection and preservation, in relation to: soil, herbage, species or the environment.

Conservation headland A sterile strip, minimum of 1m wide, between the crop and the field boundry which prevents weeds such as sterile brome and couch encroaching on the crop. A band of about 6m wide along the outside of the crop is left unsprayed and receives no pesticide treatments in the spring.

Conservation tillage Any tillage sequence, the object of which is to minimize or reduce loss of soil and water; operationally, a tillage or tillage and planting combination which leaves a 30% or greater cover of crop residue on the surface.

Consistence 1. Properties of soil material that determines its ability to change shape.

2. Consistency.

Consistency The resistance of a soil to deformation or rupture.

Consolidate To increase the density and reduce pore space in soil.

Conspecific Belonging to the same species.

Constitutive Constantly present, whether there is demand or not.

Consumer Organisms that get their energy from eating other plants and animals.

Consumptive use 1. The water used by plants.

2. Vapor loss. Expressed as equivalent depth of free water per unit of time.

Contact fungicide (protectant fungicide) A fungicide that remains on the surface where it is applied; no after-infection activity .

Contact herbicide A herbicide that is not transiocated; the activity only occurs within the first few cells from the point of contact; an example would be paraquat; a "burner."

Contact inhibition The inhibition of movement or proliferation of normal cells that results from cell-cell contact.

Contact Process resulting in two objects (i.e. herbicide molecule and a plant) touching each other.

Contact" herbicide A herbicide that does not move from the point of initial uptake.

Continuous Not deviating from uniformity; the opposite of interrupted.

Continuous cropping Growing the same crop in the same location repeatedly.

Contour An imaginary line connecting points of equal elevation on the soil's surface.

Contour basin A basin made by a levee or border built on a contour.

Contour plowing The plowman keeps to a level line at a right angle to the direction of the slope.

Contour terrace Leveling soil out on a slope at a right angle to the direction of the slope.

Contract growing Crops grown under contract to supply a specific outlet or to a factory for further process. Often grown to a detailed specification.

Contractile ring A structure of actin and myosin II that forms beneath the plasma membrane during mitosis and mediates cytokinesis.

Contractile vacuole In many protists, a specialized vacuole with associated channels designed to collect excess water in the cell. Microtubules periodically contract to force this excess water out of the cell, regulating the cell's osmotic balance.

Control A treatment that reproduces all aspects of an experiment except the variable of interest. Most well-designed experiments have control and treatment groups. An experimental manipulation is applied to the treatment group, and nothing is done to the control group. For example, if you are interested in looking at how salt water affects the growth of corn seedlings, you would design an experiment with "control" corn seedlings and "treatment" corn seedlings. You would water the treatment seedlings with salty water, and the control seedlings with non-salty water. There should be no other differences between control and treatment groups - they both should have the same amount of sunlight, should receive the same quantity of water, should be exposed to the same temperature, and so forth. If, after your salt water experiment, there are differences in corn growth (for example, perhaps the corn with salt water grew less quickly), you can confidently explain that the differences were caused by the salt. If there were two different variables between the control and treatment groups (for example, salt water and different amounts of sunlight), you would not be able to tell what made the corn grow less quickly - the salt, or the lack of sun.

Control point An enzymatic reaction that influences the flux through a pathway.

Controlled Atmosphere Regulation of O_2 and CO_2 levels as well as temperature to improve storage of fruit and vegetables

Convection The circulatory motion that occurs in fluids due to differences in temperature, which cause variations in density.

Conventional tillage Primary and secondary tillage operations normally performed in preparing a seedbed and/or cultivating for a given crop grown in a given geographical area, usually resulting in <30% cover of crop residues remaining on the surface after completion of the tillage sequence.

Converging plate boundary Boundary along which two lithospheric plates are moving towards each other; e.g. the Himalayas are forming as two plates collide.

Convolute Rolled together, or one part over another. Specifically applied to leaves or petals rolled up lengthwise in the bud.

Copper An element used in various forms, including copper sulfate and Bordeaux mixture, to control fungal and bacterial diseases of plants. Also used in strips around tree trunks to act as a barrier to snails.

Coppice Regrowth of woody plants by sprouts from stumps or roots.

Coppice woodland Consists of shoots arising from stumps of trees which are cut on a regular cycle. Used to supply small diameter poles for building, fencing and fuel. Creates a variable habitat.

Coral Colonial animal that secretes a hard outer calcareous (calcium carbonate) skeleton.

Coralloid root A root of plants in some Cycadophyta that is formed at the surface of the soil and that contains symbiotic cyanobacteria that provide nitrogen fixation.

Cordate Heart-shaped (leaf base).

Coremium Compact or fused, generally upright conidiophores, with branches and spores forming a headlike cluster.

Coriaceous Leathery.

Cork The external protective tissue of a stem or root. It is made up of polygonal-shaped cells that are non-living at maturity. The cell walls of cork tissue contain a fatty material (suberin) that makes them impermeable to water and gasses. Cork cells form the outermost layer of bark.

Corking Gradual ripening of a stem from the base, to produce a bark-like texture.

Corm A corm in a rounded, thick underground stem that is modified into a mass of storage tissue. The foliage and flowers grow from buds on the stem. Corms are usually round and slightly flattened, not pointed like a true bulb. If you cut one in half, you won't see scale rings. (See Bulb to distinguish.) The tunics of corms are usually netted and fibrous. One or more growing points, or eyes, grow from the top of the corm. At the bottom, roots grow from a basal plate. As the plant grows, the old corm dies, and new corms, called cormels, form around it. If a new cormel is large, it may produce flowers the following year, but normally it takes two to three years. Crocuses, montbretia, and gladioluses are typical corms.

Cormel A smaller, secondary corm produced by an old corm; a bulblet.

Corms A thick, rounded underground stem that is modified into storage tissue; ex. crocus, gladiolus

Corn belt An agricultural region of the central United States primarily in Iowa and Illinois, but also parts of Indiana, Minnesota, South Dakota, Nebraska, Kansas, Missouri, and Ohio.

Cornute Having horns.

Corolla The petals collectively as a unit

Corona A crown-like appendage or series of united appendages on the inner side of a corolla. It often resembles an outgrowth of the perianth, as in the Daffodil, or of the staminal circle, as in the milkweed.

Coronule A little crown of a seed; the downy tuft on seeds.

Correlation The process of determining the relationship between plant nutrient uptake or yield and the amount of nutrient extracted by a particular soil test method.

Cortex Region of parenchyma tissue between the epidermis and the phloem in stems and roots; region beneath the rind of a sclerotium.

Corticosteroids Steroid hormones produced by the adrenal gland.

Corymb A flat-topped or dome-shaped flower cluster with the outer flowers opening first, as in species of the group Achillea.

Corymbose Having flowers in corymbs.

Cosmid A vector that contains bacteriophage λ sequences, antibiotic resistance sequences, and an origin of replication. It can accommodate large DNA inserts of up to 45 kb.

Cost/benefit analysis An assessment of the costs and benefits associated with a process, action, enterprise or system.

Co-suppression The silencing of a gene do to introduction of a second copy of the gene.

Co-transforming Transforming a cell with two different genes at the same time.

Cotyledon A seed leaf present as part of an embryonic seed plant sporophyte. Cotyledons absorb nutrients from the female gametophyte or endosperm, and in some species are also photosynthetic after germination of the seed.

Cover crop A crop of nitrogen fixing plants that is planted with, or in rotation with, edible crops to increase nitrogen, stabilize soil, attract beneficial insects, or to be used as green manure.

Crease The longitudinal furrow on the ventral side of the wheat grain.

Creep soil The downward, slow, irregular mass movement of sloping soil.

Creeper A plant that grows along the surface of the soil or any other surface such as a wall or fence, by sending out rootlets from the stem. (i.e. Ivy -Hedera, Virginia Creeper - Parthenocissus quinquefolia, Trumpet Creeper - Campsis radicans)

Crenate Toothed with shallow, rounded scallops, as the edges of a leaf. When the scallops have smaller ones upon them, the leaf is said to be doubly crenate.

Crenulate Minutely crenate.

Crest Also called fasciation; a mutation that results when the growing point of a plant forms a long line, rather that a single point. In botanical terms, it is usually signified by cristate.

Cretaceous The last period of the Mesozoic, approximately 144 to 65 million years ago.

Crispate Curled or rippled at the edge such as the leaves of Pittosporum eugenioides. Also crisped.

Crista A fold in the inner mitochondrial membrane extending into the matrix.

Critical point The point where soil test values delineate responsive soils (those where fertilizer additions increase yields) and nonresponsive soils.

Crocks Broken pieces of clay pot, used to cover drainage holes of pots in order to provide free drainage and air circulation to the root system and to prevent the medium (i.e. soil) from washing down the holes.

Crop rotation Crop rotation is the practice of growing a series of dissimilar type of crops in the same space in sequential seasons to avoid the buildup of pathogens and pests that often occurs when one species is continuously cropped. Crop rotation also seeks to balance the fertility demands of various crops to avoid excessive depletion of soil nutrients. A traditional component of crop rotation is the replenishment of nitrogen through the use of legumes in sequence with cereals and other crops. It is one component of polyculture. Crop rotation can also improve soil structure and fertility by alternating deep-rooted and shallow-rooted plants.

Crop rotation avoids a decrease in soil fertility, as growing the same crop repeatedly in the same place eventually depletes the soil of various nutrients. A crop that leaches the soil of one kind of nutrient is followed during the next growing season by a dissimilar crop that returns that nutrient to the soil or draws a different ratio of nutrients, for example, rice followed by cotton. By crop rotation farmers can keep their fields under continuous production, without the need to let them lie fallow, and reducing the need for artificial fertilizers, both of which can be expensive.

Legumes, plants of the family Fabaceae, for instance, have nodules on their roots which contain nitrogen-fixing bacteria. It therefore makes good sense agriculturally to alternate them with cereals (family Poaceae) and other plants that require nitrates. A common modern crop rotation is alternating soybeans and maize (corn). In subsistence farming, it also makes good nutritional sense to grow beans and grain at the same time in different fields.

Crop rotation is also used to control pests and diseases that can become established in the soil over time. Plants within the same taxonomic family tend to have similar pests and pathogens. By regularly changing the planting location, the pest cycles can be broken or limited. For example, root-knot nematode is a serious problem for some plants in warm climates and sandy soils, where it slowly builds up to high levels in the soil, and can severely damage plant productivity by cutting off circulation from the plant roots. Growing a crop that is not a host for root-knot nematode for one season greatly reduces the level of the nematode in the soil, thus making it possible to grow a susceptible crop the following season without needing soil fumigation.

It is also difficult to control weeds which are similar to the crop, and which may contaminate the final produce. For instance, ergot in weed grasses is difficult to separate from harvested grain. A different crop allows the weeds to be eliminated, breaking the ergot cycle.

This principle is of particular use in organic farming, where pest control may be achieved without synthetic pesticides. A general effect of crop rotation is that there is a geographic mixing of crops, which can slow the spread of pests and diseases during the growing season. The different crops can also reduce the effects of adverse weather for the individual farmer and, by requiring planting and harvest at different times, allow more land to be farmed with the same amount of machinery and labour.

The choice and sequence of rotation crops depends on the nature of the soil, the climate, and precipitation which together determine the type of plants that may be cultivated. Other important aspects of farming such as crop marketing and economic variables must also be considered when choosing a crop rotation.

Crop rotation was already mentioned in the Roman literature, and referred to by great civilizations in Africa and Asia. From the end of the Middle Ages until the 20th century, the three-year rotation was practised by farmers in Europe with a rotation: rye or winter wheat, followed by spring oats or barley, then letting the soil rest (fallow) during the third stage. The fact that suitable rotations made it possible to restore or to maintain a productive soil has long been recognized by spring crops, in place of grains for human consumption.

A four-field rotation was pioneered by Flemish farmers, namely in the region Waasland in the early 16th century and popularised by the British agriculturist Charles Townshend in the 18th century. The system (wheat, barley, turnips and clover), opened up a fodder crop and grazing crop allowing livestock to be bred year-round. The four field rotation was a key development in the British Agricultural Revolution.

Crop rotation was pioneered in the United States by George Washington Carver, not only through laboratory work, but through practical promotion and education.

In the Green revolution, the practice of crop rotation gave way in some parts of the world to the practice of simply adding the necessary chemical inputs to the depleted soil, e.g., replacing organic nitrogen with ammonium nitrate or urea and restoring soil pH with lime. However, disadvantages of monoculture from the standpoint of sustainable agriculture have since become apparent.

Cross 1. To produce a hybrid plant by cross-fertilizing individuals of different varieties or species.

2. A hybrid produced by cross-fertilization.

Cross breeding Breeding from a male and female of different lineage.

Cross cells These cells are only found in grasses. They are formed from cells of the pericarp that elongate transversely, lose their chlorophyll and become lignified. They lie between the tube cells and the parenchyma of the pericarp.

Cross-fertilization A process in which sexual reproduction occurs as a result of the fusion of sex cells from different individuals .

Crossing The deliberate mating of two parental types of organisms in genetic analysis.

Crossing-over Exchange of parts between two paired chromosomes during meiosis, resulting in new combinations of linked genes within the resulting haploid cells.

Crossover events A crossover event is the exchange between two DNA molecules. The resultant DNA contains sequences (hence genes) not found in either parental DNA molecule. Each exchange is an event. Because the exchange occurs between strands of DNA, the genes are physically linked. The maxiumum number of crossover events between two genes is 50% (which is identical to independent assortment).

Cross pollination Transfer of pollen from the anthers to the stigma of another flower .

Cross protection The process whereby a normally susceptible host is infected with a less virulent pathogen (usually a virus) and thereby becomes resistant to infection by a second, usually related, more virulent pathogen.

Cross resistance Cross-resistance is resistance within a plant to herbicides with the same site of action but from different herbicide families.

Crown Upper dome of tree, bearing leaves, flowers, and fruits; junction of root and stem of a plant, usually at the soil line; in grafted woody plants, the rootstock portion of the plant near the soil surface.

Crozier The hook of an ascogenous hypha before ascus development; the curved apical portion of a blighted stem.

Cruciate Having leaves or flowers in the shape of a cross with equal arms, as certain members of the Mustard family.

Crumb structures The porous granular structure in soil.

Crust A thin, dry, hard layer of soil that forms on the surface of many soils when they are exposed to excessive heat.

Crustaceous Having a hard, brittle texture.

Crustose a form of lichen in which the thallus is thin and closely appressed to the substrate.

Cry proteins Protein crystals produced naturally by Bacillus thuringiensis bacteria that are toxic to some insects. The gene that produces this protein can be inserted into the corn genome giving it European corn borer (ECB) resistance.

Cryptobiosis (hidden life) The ability to enter a state of suspended metabolic activity during unfavorable environmental conditions (drying, cold); survival strategy of some plant pathogenic nematodes .

Cryptogam A plant, such as a fern, which reproduces by means of spores rather than seeds.

Crystalline wax Crystalline looking wax occurring on leaf surfaces that is difficult to wet.

CTD (conductivity/temperature/depth) One of the most commonly used electronic instruments in oceanography, which is lowered on a conducting cable from a ship to measure the temperature, pressure and electrical conductivity of ocean water and the height above the seafloor. From these measurements, scientists can calculate water depth, salinity, and density. By letting cable out or pulling it in, scientists can measure water properties at different depths. Some CTDs have batteries and computers that allow them to be self-recording so they can be moored in one spot to make measurements for days or months.

CTP (Chloroplast Transit Peptide) Specific amino acid sequences added to proteins which direct them to go to the chloroplast of the plant cell.

Cullion An orchid.

Culm The jointed and usually hollow stem of grasses. It is in most cases herbaceous but is woody in the bamboo and some other stout species. The term is also sometimes applied to the solid, jointless stems of sedges.

Cultigen A cultivated plant or group of specific rank for which a wild ancestor isn't known. (i.e. Zea Mays)

Cultivar A contraction of "cultivated variety" (abbreviated to cv); a group (or one among such a group) of cultivated plants clearly distinguished by one or more characteristics and which retains these characteristics when propagated; a distinct variety or race of plants that originated under cultivation and persists under cultivation. A cultivar may or may not be referable to a botanical species. Cultivars are given a name, usually distinguished by the use of single quotation marks, as in Zinnia elegans 'Tom Thumb'. Named varieties of hybrid Tea Roses and of garden Apples are examples of cultivars.

Cultivations Working of the soil prior to and during the germination of crops. For seed bed preparation and weed control

Cultural practices The manner in which plants are grown, such as: application of nutrients, irrigation practices, type of cultivation; may be used for disease management.

Culture Growth and propagation of microorganisms on nutrient media; growth and propagation of living plants.

Cuneate Wedge-shaped.

Cupule A cup-shaped involucre, occurring especially in the Oak, Beech, Chestnut and Hazel, formed by fused bracts.

Curl Distoration, puffing, and crinkling of a leaf resulting from unequal growth.

Cusp A sharp, stiff point.

Cuspidate Ending in a rigid point. An example is a leaf that is tipped with a sharp point or spine, such as the leaves of thistles.

Cut and come again Any plant that is cut or sheared after flowering and blooms again (i.e. Petunia, Pansy). Said of plants that continue to bloom when flowers are removed or cut.

Cut back To prune.

Cut flowers Flowers that are cut off the plant and used as decoration, usually indoors in a vase of water along with other flowers.

Cuticle The waxy or fatty layer covering epidermal cells of plant surfaces exposed to air such as leaves, stems and fruit. The cuticle is water-repellent and aids the plant in conserving water by reducing the amount of water vapour lost to the air from plant surfaces, particularly the upper surface of leaves.In zoology, cuticle also refers to the non-cellular outer layer of an insect or a nematode.

Cutting A section of a plant that is removed and used for propagation. A cutting may consist of a whole or part of a stem (leafy or nonleafy), leaf, bulb or root. A root cutting consists of root only; other cuttings have no roots at the time they are made and inserted. A propagation that consists of part of the crown of a plant or of its aboveground portion and roots is called a division (go see).

CV (abbr. for cultivar; variety) A plant type within a species, resulting from deliberate manipulation, which has recognizable characteristics (color, shape of flowers, fruits, seeds and height or form).

Cyanobacteria The largest and most complex prokaryotes in which photosynthesis is believed to have evolved.

Cyanobacterium (cyanobacteria) A photosynthetic prokaryote of the Division Cyanophyta.

Cyathiform Cup-shaped.

Cyathium A type of flower characteristic of Poinsettia.

Cyathus A small conical or cup-shaped organ or cavity of a plant.

Cycle A closed circle or whorl of leaves.

Cyclic Arranged in circles or whorls.

Cyclic AMP (cAMP) Adenosine monophosphate in which the phosphate group is covalently bound to both the 3′ and 5′ carbon atoms, forming a cyclic structure; an important second messenger in the response of cells to a variety of hormones.

Cyclic Electron Flow An electron transport pathway associated with photosystem I that produces ATP without the synthesis of NADPH.

Cyclic GMP (cGMP) Guanosine monophosphate in which the phosphate group is covalently bound to both the 3′ and 5′ carbon atoms, forming a cyclic structure; an important second messenger in the response of cells to a variety of hormones and in vision.

Cyclins A family of proteins that regulate the activity of Cdks and control progression through the cell cycle.

Cyclohexanediones Family of herbicides that belong to the lipid synthesis inhibitor mode of action (dims).

Cyclospermous Having the embryo coiled about the central albumen, as the seeds of Caryophyllaceae.

Cylasm A compartment in a plant or animal cell surrounding the nucleus in which many cell organelles and molecules are suspended.

Cylasmic cell The smaller, cylasm-rich cell from the cell division of a progenitor cell. Also called embryogenic cell. It will develop into either proembryonal complex or a somatic embryo.

Cyme Flat-topped or dome-shaped flower head with the inner flowers opening first. (i.e. buttercup and anemone)

Cymose Bearing a cyme; resembling or pertaining to a cyme.

CYST In fungi, a resting structure in a protective membrane or shell-like enclosure;

in nematode females, the egg-laden carcass of a female nematode;

in bacteria, a specialized type of bacterial cell enclosed in a thick wall, often dormant and resistant to environmental conditions.

Cysteine Nonessential amino acid (essential in infants and those with chronic diseases)

Cytochalasin A drug that blocks the elongation of actin filaments.

Cytochrome An iron containing pigment that plays a major role in respiration.

Cytochrome oxidase A protein complex in the electron transport chain that accepts electrons from cytochrome c and transfers them to O_2.

Cytochrome P450 Also known as cytochrome P-450 monooxygenases or CytP450s; multifunctional enzymes widely distributed among living organisms. They function as monooxygenases by binding and activating molecular oxygen, incorporating one of its atoms into an organic substrate, and reducing the second atom to form water.

Cytokines Growth factors that regulate blood cells and lymphocytes.

Cytokinesis (cytokineses) Cytoplasmic division in mitosis or meiosis.

Cytokinins Plant hormones (growth regulators) that control cell division and are important for shoot stimulation of callus in tissue culture.

Cytology The study of cells.

Cytopathology The study of changes induced by disease at the cellular level.

Cytoplasm A compartment in a plant or animal cell surrounding the nucleus in which many cell organelles and molecules are suspended.

Cytoplasmic cell The smaller, cytoplasm-rich cell from the cell division of a progenitor cell. Also called embryogenic cell. It will develop into either proembryonal complex or a somatic embryo.

Cytoplasmic inheritance (also extrachromosomal or maternal inheritance) Inheritance of genes not located in the nucleus, i.e. those in mitochondria and chloroplasts.

Cytosine A pyrimidine that base-pairs with guanine.

Cytoskeleton Integrated system of molecules within eukaryotic cells which provides them with shape, internal spatial organization, motility, and may assist in communication with other cells and the environment. Red blood cells, for instance, would be spherical instead of flat if it were not for their cytoskeleton.

Cytosol The liquid substance of prolasm, excluding all the organelles such as nuclei, plastids, ribosomes.

D

Dactyloid Finger-like in form or arrangement.

Damping off A common name for a soil disease caused by a number of different organisms. It causes seeds to rot or seedlings to collapse. The best ways to prevent this is by planting in sterile soil, treating seeds with a fungicide, or soaking the soil with a fungicide; also avoid over-watering and planting warm-season vegetables in cold soil.

Dasyphyllous 1. Having woolly leaves.

2. Having thickly set leaves.

Days to harvest Term for the prescribed minimum number of days required by U.S. government regulations between a pesticide application and the harvest of the crop.

Deadheading The removal of dead flowers or flower heads.

Dealkalization The removal of exchangeable sodium from the soil by a process of leaching or chemically treating the soil.

Decay The gradual decomposition of organic matter.

Decayed organic matter Decomposed plant or animal matter, used to improve the soil.

Deciduous Losing its leaves annually at the end of the growing season; semi-deciduous plants lose only some leaves.

Declinate Bent or curved downward or forward.

Decompose To release chemicals from organic matter by means of bacterial action.

Decomposition Breakdown into constituent elements through chemical change.

Decompound Divided into a number of compound divisions, such as a leaf divided into leaflets or a cluster of flowers divided into numerous branchlets.

Decoy crop Crop that stimulates germination of seeds of a parasitic plant such as witchweed (Striga spp.), but is not susceptible to infection by the parasitic plant; helps reduce seed populations of the parasite in soil so a susceptible crop can be planted.

Decumbent Reclining with the tips ascending; a plant that has its base lying on the ground and a stem that grows upward.

Decurrent Extending down the stem (i.e. a leaf with a base extending downward along the stem).

Decussate Arranged in pairs alternately crossing each other at right angles.

Dedicated fuel cropping Use of resources solely towards the production of a crop destined for use as a fuel

De-differentiation The developmental process from a highly differentiated cell to an undifferentiated cell. It is a reverse to differentiation.

Deep percolation The downward movement of water into the soil beyond the reach of a plant's roots.

Deep soil Soil that is deeper than 40 inches and continues until it becomes rock or another strongly contrasting material.

Defensins Antimicrobial proteins that inhibit the growth and development of pathogens. Plant defensins are found throughout the plant kingdom and are released upon seed germination, creating an antimicrobial environment around the seed while it germinates.

Deficiency correction Concept that a nutrient should be applied only if there is a reasonable expectation of a crop response; one of three approaches to making fertilizer recommendations; fertilizers would be applied only if they increased yields, and then only at optimum rates.

Deficiency payments Payment to producers to make up the difference between the average market price for a commodity and the guarantee price when the market falls below the guaranteed price.

Deflexed Bent backward or downward.

Deflocculate To break up the soil into individual particles.

Defluent Running downward; decurrent.

Defluvium A falling off, such as the bark of a tree, from disease.

Defoil To strip a plant of leaves.

Defoliate To deprive of foliage; to cut or pick the leaves of a plant, especially too early.

Defoliation The loss of leaves (for instance, from the attack of insects); specifically, the dropping of leaves in autumn.

Defoliator That which defoliates or strips of growth; specifically, in entomology, any insect that destroys the foliage of trees.

Deforest To deprive of forests; to cut down and clear away forests.

Deforestation The act of cutting down and clearing away forests of a region or stretch of land.

Degradation The breakdown of a chemical by either chemical or physical reactions that make it inactive.

Degree-days The departure of the average daily temperature from a defined base (e.g. the minimum recognized temperature for the growth of a plant species). The number of degree-days may be totaled to assess the accumulated warmth of a particular year's growing season.

Dehisce Split open to release something, usually spores or seeds.

Dehiscence The bursting open of a pod or capsule to release the seeds or of an anther to release pollen.

Dehiscent Opening, such as the seedpod of a plant.

Dehydration The loss of water.

Deltoid Triangular and attached to the stem at the wide end, such as a deltoid leaf.

Demicyclic A rust fungus that lacks the urediniospore (repeating) stage (e.g. many species of Gymnosporangium) .

Demography The study of the characteristics of populations, such as size, growth, density, distribution, and vital statistics.

Denaturation Step The first temperature step in PCR. During this step, the DNA is made single-stranded, denatured.

Denature A process which breaks the H-bonds between DNA bases, making the molecule become single-stranded.

Dendritic Branched in a way that resembles trees or shrubs.

Dendroid Tree-like; dendriform; arborescent; branching like a tree.

Denitrification $NO_3 \longrightarrow N_2O$ and/or N_2 Reduction of nitrogen oxides (usually nitrate and nitrite) to molecular nitrogen or nitrogen oxides with a lower oxidation state of nitrogen by bacterial activity (denitrification) or by chemical reactions involving nitrite (chemodenitrification). Nitrogen oxides are used by bacteria as terminal electron acceptors in place of oxygen in anaerobic or microaerophilic respiratory metabolism.

Denshire To improve land by burning piles of earth, turf, and stubble and then spreading the ashes over the ground as compost. Also densher.

Density gradient centrifugation A method of separating particles by centrifugation through a gradient of a dense substance, such as sucrose or cesium chloride.

Dentate Toothed with sharp teeth directed outward as in a dentate leaf, or having tooth-like projections, such as a dentate root.

Denticulate Minutely dentate.

Deoxyribonucleicacid (abbr. DNA) The double-stranded, helical molecule that contains genetic code information; each repeating unit, or nucleotide, is composed of deoxyribose (a sugar), a phosphate group, and a purine (adenine or guanine) or a pyrimidine (thymine or cytosine) base.

Deposition The processes by which chemical constituents move from the atmosphere to the earth's surface. These processes include rain, fog, and clouds (wet deposition), as well as particle and gas deposition (dry deposition). The term, "acid deposition," includes acid rain and refers to all forms of wet and dry deposition that have a lower than normal pH.

Depressed Flattened from above.

Deprotonated Dissociated; a molecule that has lost its proton due to a change in pH.

Derived character See apomorphy.

Desalination The removal of salts from sea water (generally)

Desalinization The removal of salts from saline soil.

Desert soil Light-colored surface soil that develops in warm- to cool-arid climates.

Desiccant A chemical causing drying

Desiccate To dry out.

Desilting area An area used to remove sediment from flowing water.

Desmosome A region of contact between epithelial cells at which keratin filaments are anchored to the plasma membrane.

Desorption The removal of absorbed materials from surfaces.

Detachment The removal of transportable fragments of soil material from a soil mass by an eroding agent, usually falling raindrops, running water, or wind; through detachment, soil particles or aggregates are made ready for transport.

Detailed soil map A map showing the different classifications of soil in a designated area.

Determinate Ceasing vegetative growth when the first flower or reproductive structure forms .

Determinate growth Cessation of ontogeny when an organism reaches its final adult form; characteristic of animals.

Detoxication Altered so the compound is less toxic.

Detritus Parts of dead organisms and cast-off fragments and wastes of living organisms. The forest floor is covered in detritus.

Deuteromycetes (Fungi Imperfecti) Group of fungi without a sexual stage; the asexual stage of ascomycetes and basidiomycetes.

Dewpond Small pond which forms on high chalky soils, filled by rainwater. Often man made for livestock drinking, kept full by rain.

Dextrorse Twining spirally upward around an axis, as in the hop or morning glory.

Diacylglycerol (DAG) A precursor in membrane lipid synthesis; diacylglycerol has two fatty acid groups attached to a glycerol back bone

Diadelphous Having stamens united in two sets by their filaments. In pea and bean flowers, nine out of ten stamens are usually united while one is by itself.

Diageotropic Growing at right angles to the direction of gravity.

Diageotropism A turning of the roots, branches, or rhizomes in a direction at right angles to that of gravity.

Diagnostic The investigative tools and techniques used in biological studies. For example, a particular set of diagnostic tools may be used to detect and identify microorganisms in food, water, soil or other samples.

Diaheliotropic The turning of the upper surfaces of a plant's leaves toward the sunlight; pertaining to diaheliotropism.

Diaheliotropism The tendency of a plant to turn the upper surfaces of its leaves toward the light.

Diakinesis The final stage of the prophase of meiosis I during which the chromosomes fully condense and the cell progresses to metaphase.

Dialypetalous Plants that have the corolla of the flower divided into distinct petals; also polypetalous.

Dialyphyllous Made of separate leaves. Applied to a calyx formed from several sepals or a corolla formed by several petals.

Dialysepalous Having a calyx made of separate sepals; polysepalous.

Diandrous A flower with two stamens.

Diapause A period of spontaneous dormancy, independent of environmental conditions, interrupting developmental activity in an embryo, larva or pupa.

Diatomaceous earth A material made from the skeletal remains of diatoms (single-celled algae). It is used as an insecticide. The sharp edges of the diatoms wound soft-bodied insects like ants and aphids. It is ordinarily used as a barrier but can be applied to leaves. It is most effective in hot, dry weather.

Diazinon A man-made insecticide widely used against many soil pests, especially in lawns; however, it is very toxic to birds.

Dibble A pointed tool used to make holes in the ground for seeds and seedlings.

Dichasial Pertaining to or resembling a dichasium.

Dichasium A flower cluster with two main stalks or branchlets within the cluster.

Dichlamydeous Having both a calyx and a corolla.

Dichogamous Exhibiting or characterized by dichogamy. Also dichogamic.

Dichogamy A provision in hermaphroditic flowers to ensure cross-fertilization (and prevent self-fertilization) by a difference in the time of maturity of the anthers and the stigma.

Dichotomous Branching, often successively, into two more or less equal arms.

Dichotomy A mode of branching by repeated forking, as is shown in some stems, the veining of some leaves, etc.

Diclinism The state of having stamens or pistils in separate flowers; the state of being diclinous.

Diclinous Having stamens or pistils in separate flowers.

Dicoccous Formed of two one-seeded carpels.

Dicotyledon Dicotyledons or "dicots" is a name for a group of flowering plants whose seed typically contains two embryonic leaves or cotyledons. There are some 199,350 species within this group. Flowering plants that are not dicotyledons are monocotyledons, typically having one embryonic leaf.

It is now accepted, thanks to the Angiosperm Phylogeny Group, that monocotyledons evolved from within the dicotyledons, and as such the latter form a paraphyletic group. This means that dicotyledons no longer are regarded as a "good" group, and the names "dicotyledons" and "dicots" are no longer to be used, at least in a taxonomic sense. The vast majority of the former dicots, however, form a monophyletic group called the eudicots or tricolpates. These may be distinguished from all other flowering plants by the structure of their pollen. Other dicotyledons and monocotyledons have monosulcate pollen, or forms derived from it, whereas eudicots have tricolpate pollen, or derived forms, the pollen having three or more pores set in furrows called colpi.

Traditionally the dicots have been called the Dicotyledones (or Dicotyledoneae), at any rank. If treated as a class, as in the Cronquist system, they may be called the Magnoliopsida after the type genus *Magnolia*. In some schemes, the eudicots are treated as a separate class, the Rosopsida (type genus *Rosa*), or as several separate classes. The remaining dicots (palaeodicots) may be kept in a single paraphyletic class, called Magnoliopsida, or further divided.

Dicotyledoneae One of two major classes of plants, the other being Monocotyledonae. Dicotyledons have two seed leaves; the mature leaves branching with net-like veins and the flowers have five, or sometimes, four parts. They include deciduous trees, broadleaf evergreens, flowering perennials, most vegetables, and weeds.

Dictyogen A monocotyledon having net-veined leaves.

Dictyogenous Having the character of a dictyogen; having the general character of an endogen, but with netted leaf-veins.

Dictyostelium discoideum A unicellular eukaryote used for studies of cell movement and cell-cell signaling.

Didymous Twofold; twin; growing double, such as the fruits of umbelliferous plants or the tubers of some orchids.

Didynamous Having four stamens in two unequal pairs, such as most mints.

Didynamy The condition of being in two unequal pairs, such as stamens.

Dieback Progressive death of shoots, branches and roots, sometimes even leading to complete death of the plant.

Differential host (differential cultivar) A plant host that on the basis of disease symptoms serves to distinguish between various strains or races of a given plant pathogen.

Differential medium Culture medium that is used to distinguish between organisms that can grow on it.

Differentiate Change from a simple to a more complex structure or function. Developing tissues and organs become specialized for particular functions. n. differentiation .

Differentiation The physiological and morphological changes that occur in a cell, tissue, or organ during development from a juvenile state to a mature state.

Diffuse Spreading widely and loosely, extended; dispersed; scattered.

Diffusion Process whereby molecules move from areas of higher concentration to areas of lower concentration.

Dig To break up and turn over piecemeal, as a portion of ground.

Digitate Having deep radiating divisions, like fingers, arising from one point; especially applied to compound leaflets that grow from the top of the leaf stem.

Dikaryon Having two sexually compatible haploid nuclei per cell, that divide simultaneously; this phase is called the dikaryophase.

Dikaryotic Having two different and distinct nuclei per cell; found in the fungi. A dikaryotic individual is called a dikaryon.

Dilution plating A method to obtain pure colonies of bacteria and fungi in which infected plant material or infested soil is diluted in sterilized water. Small samples of the water are spread on the medium surface of several petri plates to find which dilution will produce pure colonies.

Dilution streaking Repeated streaking of bacteria on the surface of a nutrient medium with a sterile metal loop to allow pure colonies to grow.

Dimerous Having two parts in each whorl of flowers.

Dimidiate Having one part so much smaller than the other as to appear to be missing, or almost so.

Dimorphic Having two distinct shapes or forms.

Dimorphism The occurrence of two distinct forms of flowers, leaves, or other parts on the same plant, or on plants of the same species, as in some Eucalyptus.

Dimorphous Existing in two forms of leaf or flower or other part on the same plant or on plants of the same species.

Dinitrification Conversion of nitrate to various gaseous forms of nitrogen which can be lost to the atmosphere (nitric oxide, nitrous oxide, dinitrogen); occurs under oxygen-limiting conditions such as when soils are saturated.

Dinitroanilines Family of herbicides that inhibit microtubule assembly and belong to the seedling growth inhibitor mode of action (DNA).

Dioeciopolygamous Polygamous with a tendency to dioeciousness, or to the prevalence of flowers of one sex on individual plants, such as a plant with mostly female flowers.

Dioecious Bearing male and female flowers on separate plants, such as in holly and willow. Also dioecian.

Dipetalous Having two petals.

Diphenylether Family of herbicides belonging to the cell membrane disrupter mode of action.

Diphyllous Having two leaves.

Diploid A condition in which the chromosomes in the nucleus of a cell exist as pairs, one set being derived from the female parent and the other from the male. In cereals such as barley and rice all the cells except the reproductive cells are diploid. When gametes are formed the diploid number (2n) is reduced by half to the haploid number (n) by meiosis When gametes fuse at fertilization the diploid condition is restored. In modern wheat the situation is more complex because it contains three genomes derived from different ancient diploid ancestors.

Diplotegia A seed capsule that develops from the ovary in some plants such as the iris, in which the ovary in situated below the floral envelopes.

Diplotene The stage of mieosis I during which homologous chromosomes separate along their length but remain associated at chiasmata.

Dipole A molecule or a functional group with a dipole moment

Dipole moment An electron asymmetry produces a local partial negative and partial charge. This separation of weak charge can be felt by the surrounding molecules. Fully charged functional groups (e.g. phosphate) also contain a dipole moment

Dipterous Having two wing-like appendages.

Direct penetration Penetration of plant tissues by a pathogen through barriers such as leaf cuticle by chemical and physical means (e.g. penetration peg).

Disboscation The act of disforesting; the act of converting woodland into farmland.

Disbud To remove buds or shoots; to remove unnecessary buds of a tree, vine or flowering plant to encourage production of high-quality flowers or fruits. Also called pinching back or stopping.

Disc A type of minimum tillage that breaks up the organic matter and partially buries it in the soil. Depth of soil disturbance is seldom over 4-5 inches.

Discoid 1. Pertaining to the button-like center of flowers belonging to the Compositae family (i.e. daisy, sunflower, etc.).

2. Having only the central part and not the outer petal-like parts of the flower, as in Tansy.

Discomycetes A group of the ascomycetes in which the hymenium is exposed at maturity; one in which the fruiting body is an apothecium or discocarp.

Disease Any malfunctioning of host cells and tissues that results from continuous irritation by a pathogenic agent or environmental factor which leads to the development of symptoms; - abnormal functioning of physiological processes of an organism.

Disease cycle Succession of all of events and interactions among the host, parasite and environment that occur in a disease, from initial infection of the plant by a causal agent, through pathogenesis, to over-seasoning, until another infection occurs.

Disease incidence Number of plants affected by a disease within a population.

Disease progress curve Graph of some measure of disease (i.e. severity) over time .

Disease pyramid A memory aid similar to the disease triangle but including, in addition, the factor of time in the development of a disease.

Disease severity The measure of damage done by a disease.

Disease triangle A memory aid that diagrams the three important components necessary for disease: susceptible plant, virulent pathogen and favorable environment.

Dish garden A collection of plants, generally small ornamentals, grown together in one container.

Disinfect To eliminate a pathogen from infected plant tissues.

Disinfection The destruction of germs of infectious diseases.

Disinfest To kill pathogens that have not yet initiated disease, or other contaminating microorganisms, that occur in or on inanimate objects as such soil or tools, or that occur on the surface of plant parts such as seed.

Disk 1. Any flattened circular organ of a plant, such as the adhesive ends of the tendrils of the Virginia creeper.

2. The button-like center of the flowers of plants such as sunflowers and asters.

3. A fleshy enlargement of the receptacle in some flowers.

Disk flower In the family Compositae, one of the miniature, tubular components of the central area of the flower head.

Dispermous Having two seeds.

Dispersal (dissemination) Spread of infectious material (inoculum) from diseased to healthy plants.

Dispersion of soil The breaking up of soil into very fine particles and its suspension in water.

Dissected Deeply divided into many narrow segments.

Dissemination (dispersal) Spread of infectious material (inoculum) from diseased plants to healthy plants.

Dissepiment A partition within an organ of a plant (i.e. the membrane that separates sections of the orange and other citrus fruits).

Dissilient Bursting open with some force, as the pod or capsule of Impatiens.

Dissociate (dissociation) The splitting of a single chemical species into two ions, for example H_2O to produce H+ and OH-.

Dissolved inorganic carbon (DIC) All inorganic carbon (e.g., carbon dioxide) dissolved in a given volume of water at a particular temperature and pressure.

Dissolved organic carbon All organic carbon (e.g., compounds such as acids and sugars, leached from soils, excreted from roots, etc) dissolved in a given volume of water at a particular temperature and pressure.

Dissolved oxygen Amount of oxygen gas dissolved in a given volume of water at a particular temperature and pressure.

Distal Far from the point of attachment or origin; in a direction away from main body .

Distichous Arranged in two vertical rows on opposite sides of an axis.

Disturbance A natural or human-induced disruption or alteration of an ecosystem. Forest fires, tornadoes, or rock slides are examples of natural disturbances; logging, acid rain, and road-building are examples of human disturbances.

Divaricate Spreading far apart.

Divarication The act of branching off or diverging; separation into branches.

Divergent Spreading.

Diverging plate boundary A boundary between two plates that are moving apart, where volcanic activity is creating new oceanic crust - the mid-oceaen ridge is an example.

Divided Separated to the base or midrib (i.e. leaves of palms or clover).

Division A method of propagation by which a plant clump is lifted and divided into separate pieces, which include roots and a growing point, during dormancy. The pieces are immediately replanted.

DNA extraction The process of isolating the DNA from cells of an organism to allow scientists to work with and study it further.

DNA markers The developmentalDNA sequences that are found at specific loci in species that may vary in sequence among members of that species.

DNA glycosylase A DNA repair enzyme that cleaves the bond linking a purine or pyrimidine to the deoxyribose of the backbone of a DNA molecule.

DNA ligase An enzyme that seals breaks in DNA strands.

DNA markers DNA sequences that are found at specific loci in species that may vary in sequence among members of that species.

DNA polymerase The enzyme which catalyzes the reaction of adding new DNA bases to a growing DNA strand.

DNA replication The process in which a DNA molecule is copied. This takes place in PCR and naturally in cell division.

DNA Sequence A sequence of nucleotides that can be from any DNA region, such as a gene or chromosome.

DNA template That particular portion of a DNA molecule which is copied in PCR.

Dodder Parasitic seed plant (Cuscuta spp.) without leaves; a yellow filamentous vine.

Dodecamerous Having the whorls of the flower (petals, sepals, etc.) in twelves. Also written 12- merous.

Dolabriform Having the form of an axhead.

Dolichol phosphate A lipid molecule in the endoplasmic reticulum upon which oligosaccharides are assembled for the glycosylation of proteins.

Dolipore septum Cross wall found in basidiomycetes and characterized by special swellings and membranes in association with the septal pore.

Dolomitic limestone A fertilizer that contains magnesium.

Domain A discrete portion of a protein with its own function. The combination of domains in a single protein determines its overall function.

Dominant A gene is said to be dominant if it expresses its phenotype (physical characteristic) even when contributed by only one of the parents.

Dominant inhibitory mutant A mutant that interferes with the function of the normal allele of the gene.

Donor parent The parent from which one or a few genes are transferred to the recurrent parent in backcross breeding.

Dormancy A resting period in the life of a plant during which growth slows or appears to stop. In cereals this is between the start of seed dry-down and maturity. A period of 'after-ripening' is often required before dormancy can be successfully broken.

Dormant spray Any type of pesticide applied to dormant, usually leafless, plants to control a number of insects and disease.

Dorsal Pertaining to or attached to the back of a plant or plant part; facing away from or situated outside the axis of a plant.

Dorsifixed Attached at the back of a plant or plant part, such as the anthers of lilies.

Dorsiventral Flattened and having a definite front and back surface, such as a blade of grass.

Double This is used to describe flowers which have more than one row of petals; this may be because the stamens and sometimes the pistils are transformed into petals or sometimes the petals split to form several more. Semi-double flowers are those which have only partially changed. Completely double flowers have lost their reproductive organs and are therefore unable to produce seeds. These flowers usually stay fresh longer than single flowers when cut. Double flowers are seen in the wild once in a while, but are mostly developed varieties.

Double fertilization The unique process, in flowering plants, when two male nuclei, from pollen, fuse separately with different female nuclei. The first male nucleus fuses with the egg cell to form the zygote that develops into the embryo The second male nucleus fuses with the two polar nuclei (the second polar event) to form a triploid nucleus that then develops into the endosperm

Double haploidy technology (microspore embryogenesis) Haploid plants are derived from microspores, or immature pollen grains, and therefore, have only one set of chromosomes. The chromosome number in these plants is doubled resulting in homozygous, pure lines. This technology allows researchers to develop true breeding lines that are easily selected for breeding. It also decreases the time taken to breed plants for particular traits.

Double helix The spiral shape that two linear strands of DNA assume when bonded together.

Double membrane In mitochondria and plastids, there is a two-layered membrane which surrounds the organelle. This is believed to be the result of endosymbiosis, with the outer membrane coming from the eukaryotic cell, and the inner membrane belonging to the original prokaryote which was "swallowed".

Double ridge This is the term used to describe the transition from the 'vegetative' to the 'floral' state in the shoot apex of wheat and barley. Growth of the leaf primordium (ridge) is suppressed and the floral bud primordium (ridge) begins to grow out, giving a 'double ridge' appearance to the dissected shoot apex when view under the microscope. At this time the plant apex no longer produces leaves but is 'florally initiated' and only flower parts will subsequently develop.

Doubly serrate Alternating large and small teeth.

Downy Covered with soft hairs or down.

Downy Mildew White or grey 'bloom' on leaves and stems caused by production of sporangiophores and sporangia by members of the Peronosporales (downy mildew fungi).

DPA Days Post Anthesis' or the number of days after flowering and fertilisation

Drain To remove water from the soil by artificial means, such as drainage ditches, buried perforated plastic pipes, or a gravel sump. Also to improve heavy clay soils with sand or organic matter.

Drainage 1. The removal of excessive water from the ground.

2. Water moving down in the soil.

3. A system of conduits, channels or passages by means of which something is drained.

Drain-well A pit sunk through an impenetrable layer of earth or stone to a porous layer, to draw off water.

Drepaniform Sickle-shaped.

Dried blood An organic fertilizer that leaves an acid reaction.

Dried flowers Blossoms that have been dehydrated but have retained a natural shape and color.

Drift (of pesticides) Movement of airborne particles of a spray, dust, or vapor away from the target area during or shortly after an application.

Drill 1. A machine used for planting seeds, as of grasses, wheat, oats, corn, etc. by dropping them in rows and covering them with earth.

2. A row of seeds deposited in the earth; the trench or channel in which the seeds are deposited.

3. To sow seeds in rows (i.e. the field was drilled, not sown broadcast).

Drip irrigation A system of watering by which moisture running through a porous hose is slowly released through tiny holes, or emitters, to the root zone of plants.

Drosophila melanogaster A species of fruit fly commonly used for studies of animal genetics and development.

Drought Insufficient soil water for normal plant growth.

Drought-tolerant plants Plants that can survive periods of time with little or no water.

Drowning A surplus of water that drowns plant roots by eliminating oxygen from the soil.

Drupaceous Pertaining to or producing drupes.

Drupe A stone fruit having a hard, nut-like inner part surrounded by a fleshy or fibrous outer layer (i.e. cherry, peach, olive, walnut); a small fruit with a single seed.

Drupelet A small drupe, especially as applied to the individual components of aggregate fruits such as raspberries and blackberries, which consist of numerous seeds, each surrounded by its own fleshy layer. Also drupel.

Druxy Pertaining to a tree or timber, having decayed spots or streaks in the heartwood.

Dry sand A sandy deposit with low water-holding capacity.

Duff The partly decomposed organic surface layer of forested soil.

Dune A mound of loose sand piled up by wind.

Durable resistance Resistance that remains effective during prolonged and widespread use in an environment favorable to disease.

Dust mulch A dry, loose surface layer of cultivated soil, which reduces water loss from the underlying soil.

Dust/Dusting Method of pesticide control in which a dry substance is applied by spraying.

Duster A device consisting of a bin, a wand with a nozzle, and a crank mechanism; used for applying dust or powder to plants.

Dwarf fruit trees 1. A small fruit tree reaching a height at maturity of 4 or 5 feet and bearing early, but normal fruit.

Dwarf plants Plants of genetically small size or those dwarfed by grafting (i.e. dwarf apple trees) or by pruning and restricted root growth (i.e. bonsai).

Dwarf shrubs Shrubs that are selected or hybridized that are typically shorter and more compact than the species.

Dwarfing Underdevelopment of a plant or plant organs, which may be caused by disease, inadequate nutrition, or unfavorable environmental conditions.

Dyeing flowers Blossoms capable of producing dyes.

Dynamic Something that changes or moves; not static or stationary.

Dynein A motor protein that moves along microtubules towards the minus end.

E

E. coli (Escherichia coli) A species of bacteria used as a model for biochemistry and molecular biology.

E2F A family of transcription factors that regulate the expression of genes involved in cell cycle progression and DNA replication.

Ear emergence Main heading date of a crop. i.e for a grass sward, the date at which 50% of the inflorescences have emerged.

Ear shank The part of the ear that is attached to the stalk of the corn plant. The ear shank is a transport for water and nutrients to the ear from the plant.

Eared Having ears or awns, such as in corn or wheat.

Early milk A description the young grain. When punctured the grain contents are starting to look milky.

Earthquake Movement (shaking or sudden motion) of the Earth caused by the rapid release of strain.

East Pacific Rise A part of the mid-ocean ridge system running northeasterly from near New Zealand to just off the coast of Mexico in the Gulf of California. There is a lot of volcanic activity associated with this seafloor spreading center, as well as many hydrothermal vents.

Ebracteate Without bracts.

Ebracteolate Without small bracts.

Ecalcarate Having no spur.

Ecarinate Without a carina or keel.

Ecdysone An insect steroid hormone that triggers metamorphosis.

Echinate Prickly; covered with stiff bristles.

Echinulate Having small spines projecting from cell walls.

Ecology The study of the interactions of living organisms with one another and with their nonliving environment of matter and energy. An example of an ecological research question is, "How do weather, soil type, and topography affect the presence and growth of tree species?"

Economic threshold The pathogen density at or above which the value of crop losses (in the absence of management efforts) would exceed the cost of management practices.

Economies of scale Unit cost reductions which result from increasing total output.

Ecosystem A community of different species interacting with one another and with the chemical and physical factors making up the nonliving environment. An ecosystem can be small (for example, a pond), or very large (for example, a major river valley).

Ectoderm The outer germ layer; gives rise to tissues that include the skin and nervous system.

Ectodesmata Ectodesmata are pits in the outer wall of epidermal cells, which connect to valleys in the cuticle surface, exposing the plasma membrane. Plant metabolites leak from the ectodesmata and accumulate in the valleys on the leaf surface, providing nutrients for pathogens and saprophytes.

Ectomycorrhiza Symbiotic association between a nonpathogenic or weakly pathogenic fungus and the roots of plants with fungal hyphae between and external to root cells .

Ectoparasite A parasite that lives and feeds from the exterior of its host's cells or tissues. Compare with endoparasite.

Ectotrophic Fungal development primarily over the root surface.

Eddish Stubble; a stubble field. Also called earsh and ersh.

Edema (also oedema; intumescence) Blisters produced on leaves and other plant parts under conditions of high moisture and restricted transpiration.

Edging 1. Any nonbiodegradable material, ordinarily plastic or metal, installed in the ground to separate flower beds from lawn or other landscape areas.

2. A border; a skirting; specifically, a row of plants set along the border of a flower bed.

Ef flux Movement from the cylasm to the cell wall region; out of the cell.

Efflorescence The time or state of flowering; anthesis.

Efflorescent Blooming or flowering.

Effluent Drainage from a cattle yard, manure heap or silage clamp.

Efflux Movement from the cytoplasm to the cell wall region; out of the cell.

Egg mass Group of eggs held together by a gelatinous matrix.

Eicosanoid A class of lipids, including prostaglandins, prostacyclins, thromboxanes, and leukotrienes, that act in autocrine and paracrine signaling.

Elater 1. In liverworts (Hepatophyta) and hornworts (Anthocerophyta), thin filaments made of diploid cells that form among the meiospores and help disperse them.

2. In Equisetum (Sphenophyta), strap-like projections from the meiospore cell walls that coil and straighten with changes in humidity, aiding dipersal.

Electric hotbed A cold frame used for extending the growing season. They are usually heated by an electrical mat or heating element.

Electrochemical gradient A difference in chemical concentration and electric potential across a membrane.

Electrochemical potential Gradients of concentration and electric potential across cell membranes

Electron asymmetry The electrons that form a chemical bond between two atoms are not shared equally, resulting in one partner with a slightly negative charge (the tight sharer) and one partner with a slightly positive charge (the weaker sharer)

Electron microscope Microscope that uses a focused beam of electrons to produce a greatly enlarged image of minute objects, such as a virus, in the same way that light is used in a compound microscope.

Electron microscopy A type of microscopy that uses an electron beam to form an image. In transmission electron microscopy, a beam of electrons is passed through a specimen stained with heavy metals. In scanning electron microscopy, electrons scattered from the surface of a specimen are analyzed to generate a three-dimensional image.

Electron orbitals Electrons move about nuclei in spaces that have a characteristic shape. These orbiting shapes are called orbitals

Electron transfer The transfer of electrons between a series of components in a controlled fashion. Examples include the mitochondrial electron transfer chain involved in oxidative phosphorylation and the chloroplast electron transfer chain carrying out phohosphorylation. Electrons are transferred from molecules with a higher free energy to those with a lower free energy.

Electron transport chain A series of carriers through which electrons are transported from a higher to a lower energy state.

Electrophilic substrate A functional group that likes electrons; is seeking a negative charge.

Electrophoresis A technique which uses electricity to separate molecule fragments according to size so they can be studied.

Electrophoresis buffer A solution of water and electrolytes that will conduct current for electrophoresis.

Electrophoresis gel A gel matrix made from agarose, starch or acrylamide that is used to seperate large molecules such as proteins or DNA based on the rate they move with an electric current.

Electroporation A method of cell transformation using a pulse of electricity to open small temporary holes in the plant or bacteria cells. The foreign DNA mixed with these cells is able to pass into the cell through these holes.

Elemental sulfur Most concentrated form of sulfur; must be oxidized to the sulfate form before plants can use it. Must be finely ground to particle sizes of 80-100 mesh to be oxidized and effective during the same growing season.

Elevated meniscus A meniscus is formed when water interacts with the sides of a vessel or tube and does not form a flat surface. In an elevated meniscus, the water forms a rounded surface toward the air

Elicitor A molecule produced by the host (or pathogen) that induces a response by the pathogen (or host).

ELISA (Enzyme-Linked ImmunoSorbent Assay) A test which can be used to detect the presence of proteins encoded by a specific gene.

Elite A crop line that has many genes for good agronomic traits that result in high yields in a particular environment.

Elite experimental line A line that is very close to release as a cultivar

Elliptic Broadening at or about the center and narrowing equally toward each end.

Elongate Lengthened.

Elongation step The last temperatures step in PCR. During this step a DNA polymerase adds nucleotides to the growing DNA strand.

Eluviation The movement of moisture and nutrients from one place to another in the soil.

EMA Environmental Management in Agriculture: proprietary Software Package. Go to: Departmental Software Agriculture and Ag. botany Environmental Management in Agriculture. Takes about 60 seconds to download files.

Emarginate Having a shallow notch at the tip, such as on a leaf.

Emargination A notching of the edge, such as along the edge of a leaf.

Embryo An organism in the early stages of development, such as a young plant in the seed, or a nematode before hatching from the egg.

Embryo proper A cell mass that will develop into an embryo.

Embryo Sac A large cell that develops in the ovule of flowering plants. It contains a large number of nuclei derived by division of the megaspore nucleus. At the micropyle end is the egg nucleus and two synergids; at the opposite end are nuclei that will become the antipodals and in the centre are two polar nuclei which fuse to become the primary endosperm nucleus.

Embryonal stem (ES) cells Cells cultured from early embryos.

Embryonic axis The major part of the embryo, consisting of the shoot, mesocotyl and the root.

Embryophyte A member of the Kingdom Plantae. Having an embryonic sporophyte dependent on the gametophyte for at least the first part of its life.

Embryotega A small circular thickening near the seed scar on the seeds of some plants, such as members of the spiderwort family.

Emergence Growth of the seedling shoot through the surface of the soil.

Emersed Rising above the water, such as the leaves of many water plants.

Emulsifier A surface-active agent promoting the formation and stabilization of an emulsion; promotes the mixing of two normally unmixable liquids; an emulsifier is used to didperse an oil within water.

Emulsion A system consisting of a liquid dispersed with an emulsifier in a second but immiscible liquid.

Enantioblastic Pertaining to a plant embryo, originating at the end directly opposite the hilum.

Enation Abnormal outgrowth from the surface of a stem or leaf.

Encapsidate To cover virus nucleic acid with a protein coat.

Encyst To form a cyst or protective covering.

Endemic Native to a particular place; pertaining to a low and steady level of natural disease occurrence.

Endobiotic Describes an organism living within the cells or tissues of a host organism. Endobiotic organisms are endoparasites.

Endocarp In fruits whose pericarp consists of dissimilar layers, the innermost layer. It may be hard and stony (i.e. plum or peach pit), membranous (i.e. apple core), or fleshy (i.e. orange pulp).

Endoconidium A conidium produced inside a hypha or conidiophore.

Endocrine signaling A type of cell-cell signaling in which endocrine cells secrete hormones that are carried by the circulation to distant target cells.

Endocytosis The uptake of extracellular material in vesicles formed from the plasma membrane.

Endoderm The inner germ layer; gives rise to internal organs.

Endodermis (endodermes) A layer of cells surrounding the vascular tissue in roots, stems, or leaves of many seed plants. The endodermis is the innermost layer of the cortex.

Endogen A plant that develops from or on the inside, as most monocots.

Endogenous 1. Of or pertaining to endogens (endogenous plants).

2. Proceeding from within (endogenous growth).

Endomycorrhiza Symbiotic association between a nonpathogenic or weakly pathogenic fungus and the roots of plants in which fungal hyphae invade cortical cells of the root .

Endoparasite A parasitic organism that lives and feeds within the cells or tissues of its host. Compare with ectoparasite.

Endophyte Plant developing inside another organism; also used for endoparasitic fungi found in grass species.

Endoplasmic reticulum (ER) An extensive network of membrane-enclosed tubules and sacs involved in protein sorting and processing as well as in lipid synthesis.

Endopolyploid When the DNA in the nucleus of a cell is replicated again and again, without any cell division taking place. The DNA C value goes from 4 to 8 to 16 to 32 etc and the chromosome number (n) goes from 2 to 4 to 8 to 16 etc.

Endosome A vesicular compartment involved in the sorting and transport to lysosomes of material taken up by endocytosis.

Endosperm A nutritive material in seeds of the Anthophyta, formed from the triple fusion nucleus.

Endospore Resistant, thick-walled asexual spore formed within a bacterial or fungal cell .

Endosporic Of gametophytes, forming inside the meiospore cell wall.

Endostome The opening in the inner covering of the ovule.

Endosulfan A broad spectrum, synthetic insecticide used to control many plant pests, including whiteflies and thrips. Common trade name is Thiodan.

Endosymbiosis A symbiotic relationship in which one cell resides within a larger cell.

Endotoxin A protein that works internally as a toxin.

Energy flow The rate of energy transfer between elements of an ecological system

Energy of activation The amount of kinetic energy required for conversion of a substrate to a product.

Enhancer A transcriptional regulatory sequence that can be located at a site distant from the promoter.

Ensiform Sword-shaped; long, narrow, flat, and tapering to a point, as the leaves of many species of iris.

Enthalpy A measure of the total bond-energy (heat) in a molecule.

Entire Undivided and without teeth.

Entomophilous Applied to flowers whose structure ordinarily requires insects to pollinate them.

Entropy A measure of the degree of disorder in a system.

Environment The combinations of all the conditions external to the genome that potentially affect gene expression.

Enzyme A protein molecule that catalyzes biochemical reactions in living systems, including those found in metabolic pathways. As a catalyst, an enzyme facilitates reactions without being consumed by them. Examples include the stomach enzyme pepsin, which digests food, and cellulase, which digests the polymeric sugar cellulose into smaller units.

Enzyme-linked immunosorbent assay (acronym ELISA) A serological test in which the sensitivity of the reaction is increased by attaching an enzyme that produces a colored product to one of the reactants .

Epappose Having no tuft of bristles, a feature often found on seeds that are scattered by the wind, such as the seeds of a dandelion.

Ephemeral gully See gully erosion

Ephemeral Short-lived; existing or continuing for a short time only.

Epiblast A small flap of tissue on the side of the grass embryo axis opposite the scutellum

Epicalyx A calyx-like involucre of bracts below the true calyx, as in the mallow, or of sepals, as in the potentilla.

Epicarp The outer skin of fruits.

Epichilium The terminal part of the lip of some orchids. Also epichil or epichile.

Epicotyl The upper portion of the embryonic axis, above the cotyledons (in cereals: the scutellar node) and below the first true leaves.

Epicotyledonary Pertaining to the epicotyl; situated above the seed leaves (cotyledons).

Epidemic An increase of disease in a population; a general and serious outbreak of disease .

Epidemiology The study of factors influencing the initiation, development, and spread of infectious disease; the study of disease in populations of plants.

Epidermal cells Cells forming a protective layer on the surfaces of plants and animals.

Epidermal growth factor (EGF) A growth factor that stimulates cell proliferation.

Epidermis The outer, protective tissue of a plant. In succulents, this may bear protective features such as thick, fine hair or bloom.

Epigeal 1. A plant that grows above the surface of the soil, as most plants in the garden do.

2. A cotyledon that is forced above the ground in the process of germination, as in beans, morning glories, and others.

3. Cotyledons that appear above ground. Also epigeous, epigean, epigaean, or epigeic.

Epigenous Growing on the surface of a leaf or other plant organ (i.e. fungi grow on the surface of leaves, such as mildew).

Epigynous Referring to the calyx, corolla, and stamens that appear to be growing from the top of a ovary, but are actually fused to it, as in the cranberry.

Epinasty The downward curvature of leaves that occurs when the upper side of the petiole grows faster than the lower side.

Epipetalous Borne on the petals or the corolla of a flower; applied to stamens and to plants whose stamens are attached to the corolla.

Epiphyllous On the upper surface of a leaf.

Epiphyseal cells The meristem cells in epiphysis

Epiphysis The shoot apical meristem tissue.

Epiphyte A plant that grows on another (such as on a tree) without being parasitic and obtains nutrients from decaying leaves and other debris and moisture from the air without rooting into the soil. These plants do not obtain nutrients directly from the host plant as do parasites and saprophytes. Many Orchids, Bromeliads, Ferns and some Cacti are epiphytes.

Epiphytotic An epidemic in a plant population.

Episepalous Growing on or fused with a sepal; applied to stamens.

Episodic Occurring or appearing at usually irregular intervals

Episperm The hard external covering of a seed.

Epispore The outer coat or membrane of a spore.

Epistasis Interaction between two loci, whereby the phenotypic effect of an allele at one locus depends on which allele is present at the other locus.

Epithelial cells Cells forming sheets (epithelial tissue) that cover the surface of the body and line internal organs.

Epitope An amino acid (or other) sequence that effects formation of an antibody.

Epixylous Growing on wood.

Epruinose Having no bloom or powdery coating upon the surface of the leaf, stem or other plant part; not pruinose.

EPS (extracellular polysaccharides) Sugar polymers that contribute to the slimy appearance of bacteria probably promote colonization of plant tissues and disease development in bacteria.

EPSPS 5-enolpyruval-shikimate-3 phosphate syntheses; an enzyme essential for amino acid production that is the target site for glyphosate (Roundup).

Equilibrium centrifugation The separation of particles on the basis of density by centrifugation to equilibrium in a gradient of a dense substance.

Equitant Leaves that overlap each other to form a fan (i.e. as many varieties of Iris).

Eradication The control of plant disease by eliminating the pathogen after it is established or by eliminating all of the plants that carry the pathogen.

Ergot A disease of certain grasses and cereals, especially rye, caused by Claviceps spp.;

Ergotism A disease caused by ingestion of grain contaminated with alkaloids of ergot fungi, often called the holy fire or St. Anthony's fire in the Middle Ages.

Ericaceous Applied to acid potting mix with a pH of 6.5 or less. It is usually used for plants that cannot tolerate lime.

Ericetal Composed of or containing heaths.

Erodible A term used to describe a soil that is vulnerable to erosion.

Erose Having an irregularly jagged margin, as if gnawed; applied to a leaf or other plant part.

Erosion The wearing away of the land through the action of moving water, wind, or other geological agents.

Erosive Used to describe water that has enough force to cause erosion.

Erucic Acid The major fatty acid found in rapeseed oil used for industrial purposes. Varieties of oil with low erucic acid content such as canola oil have nutritional value.

Erumpent Bursting or erupting through the substrate surface.

Erythrocytes Red blood cells.

Escape A plant that has escaped from cultivation and naturalized more or less permanently (i.e. Queen Anne's lace is an escaped carrot).

Espalier 1. A trelliswork of various forms on which the branches of fruit trees or fruit bushes are extended horizontally, in fan shape, etc., in a single plane. The object it to provide better air circulation and sun exposure for the plants.

2. A tree or plant trained on such a trellis or system. Trees trained to espaliers aren't as liable to suffer abrupt variations of temperature as are wall trees.

3. To train on or protect by an espalier, as a tree or trees.

Espathate Lacking a spathe.

Esquamulose Lacking minute scales; not squamulose.

ESR Electron spin resonance spectroscopy

Estrogen A steroid hormone produced by the ovaries.

ESTs (Expressed Sequence Tags) DNA sequence information developed by partial sequencing of a large number of randomly selected cDNA clones. These sequence databases describe the kinds of genes and expression levels in a particular tissue.

Etagere A series of open shelves for growing and displaying plants; a tiered plant stand.

Ethidium bromide Dye molecule with a structure that binds to DNA and will fluoresce under ultraviolet light.

Ethion A synthetic insecticide effective against sucking and chewing insects, especially whiteflies and mites.

Ethylene Ethylene (or IUPAC name ethene) is the simplest alkene hydrocarbon, consisting of four hydrogen atoms and two carbon atoms connected by a double bond. Because it contains a double bond, ethylene is called an *unsaturated hydrocarbon* or an *olefin*.

The molecule cannot twist around the double bond at room temperature, and all six atoms lie in the same plane. The angle made by two carbon–hydrogen bonds in the molecule is 117°, very close to the 120° that would be predicted from ideal sp^2 hybridization.

From 1795 on, ethylene was referred to as the *olefiant gas* (oil-making gas), because it combined with chlorine to produce the *oil of the Dutch chemists* (ethylene dichloride), first synthesized in 1795 by a collaboration of four Dutch chemists.

In the mid-19th century, the suffix *-ene* (a Greek root added to the end of female names meaning "daughter of") was widely used to refer to a molecule or part thereof that contained one fewer hydrogen atoms than the word being modified. Thus, *ethylene* (C_2H_4) was the "daughter of ethyl" (C_2H_5). The name ethylene was used in this sense as early as 1852.

In 1866, the German chemist Augustus von Hofmann proposed a system of hydrocarbon nomenclature in which the suffixes -ane, -ene, -ine, -one, and -une were used to denote the hydrocarbons with 0, 2, 4, 6, and 8 fewer hydrogens than their parent alkane[1]. In this system, ethylene became *ethene*. Hofmann's system eventually became the basis for the Geneva nomenclature approved by the International Congress of Chemists in 1892, which remains at the core of the IUPAC nomenclature. However, by that time, the name ethylene was deeply entrenched, and it remains in wide use today, especially in the chemical industry.

The double bond is a region of slightly higher electron density, and most of ethylene's chemistry involves other molecules reacting with and adding across its double bond. Ethylene can react with bromine, chlorine, and other halogens, to produce halogenated hydrocarbons. It can also react with water to produce ethanol, but the rate at which this happens is very slow unless a suitable catalyst, such as phosphoric or sulfuric acid, is used. Under high pressure, and, in the presence of a catalytic metal (platinum, rhodium, nickel), hydrogen will react with ethylene.

Ethylene is produced in the petrochemical industry via steam cracking. In this process, gaseous or light liquid hydrocarbons are briefly heated to 750–950 °C, causing numerous free radical reactions to take place. Generally, in the course of these reactions, large hydrocarbons break down in to smaller ones and saturated hydrocarbons become unsaturated.

The result of this process is a complex mixture of hydrocarbons in which ethylene is one of the principal components. The mixture is separated by repeated compression and distillation.

Another process is catalytic cracking where it is used in oil refineries to crack large hydrocarbon molecules into smaller ones. Use of zeolite as a catalyst allows the cracking to be achieved at a lower temperature. It is an important way of separating alkenes from alkanes using a fractionating column.

Although ethylene is a relatively simple molecule, its spectra is considered to be one of the most difficult to explain adequately from both a theoretical and practical perspective. For this reason, it is often used as a test case in computational chemistry. Of particular note is the difficulty in characterizing the ultraviolet absorption of the molecule. Interest in the subtleties and details of the ethylene spectrum can be dated back to at least the 1950s.

Ethylene is used primarily as an intermediate in the manufacture of other chemicals, especially plastics. Ethylene may be polymerized directly to produce polyethylene (also called *polyethene* or *polythene*), the world's most widely-used plastic. Ethylene can be chlorinated to produce ethylene dichloride (1,2-Dichloroethane), a precursor to the plastic polyvinyl chloride, or combined with benzene to produce ethylbenzene, which is used in the manufacture of polystyrene, another important plastic.

Smaller amounts of ethylene are oxidized to produce chemicals including ethylene oxide, ethanol, and polyvinyl acetate.

Global demand for ethylene exceeded 100 million tonnes per year in 2005. Ethylene was once used as an inhaled anesthetic, but it has long since been replaced in this role by nonflammable gases. It has also been hypothesized that ethylene was the catalyst for utterances of the oracle at Delphi in ancient Greece. Ethylene is used in greenhouses and is sprayed on crops to speed ripening. It is also found in many lip gloss products.

Ethylene functions as a hormone in plants. It stimulates the ripening of fruit, the opening of flowers, and the abscission (or shedding) of leaves. Its biosynthesis starts from methionine with 1-aminocyclopropane-1-carboxylic acid (ACC) as a key intermediate.

"Ethylene has been used in practice since the ancient Egyptians, who would gas figs in order to stimulate ripening. The ancient Chinese would burn incense in closed rooms to enhance the ripening of pears. It was in 1864, that leaks of gas from street lights showed stunting of growth, twisting of plants, and abnormal thickening of stems (the triple response). In 1901, a Russian scientist named Dimitry Neljubow showed that the active component was ethylene. Doubt discovered that ethylene stimulated abscission in 1917. It wasn't until 1934 that Gane reported that plants synthesize ethylene. In 1935, Crocker proposed that ethylene was the plant hormone responsible for fruit ripening as well as inhibition of vegetative tissues. Ethylene is now known to have many other functions as well." - from (plant-hormones.info)

Effects

Stimulates leaf and flower senescence

Induces leaf abscission mainly in older leaves.

Induces seed germination

Induces root hair growth – this increases the efficiency of water and mineral absorption

Stimulates epinasty – leaf petiole grows out, leaf hangs down and curls into itself

Stimulates fruit ripening

Induces the growth of adventitious roots during flooding

Usually inhibits growth - although perhaps just shoot growth (speculative)

Affects neighboring individuals

Disease/wounding resistance

Triple response when applied to seedlings – root and shoot growth inhibition and pronounced hypocotyl hook bending

Inhibits stem swelling (Contradictory to the finding below – contradictory sources)

Stimulates cell broadening (and lateral root growth)

Interference with auxin transport (with high auxin concentrations)

Directly or indirectly induces auxin at high levels (speculative)

Inhibits the rate of metabolism of cells in the shoot so as to redirect resources to the root (speculative)

Is a general indicator of poor root health. Strategy of senescent leaves may to funnel more resources to the root. (speculative)

May be more active at night when root and mineral acquisition are, on average, lower (speculative)

Just as a role of auxin may be to increase minerals and water by shoot growth, ethylene may do this by shoot senescence. Cytokinin and auxin hormones are released when conditions are favorable for growth, for example during the day. Ethylene and gibberellin (or brassinosteroid) may be released when the plant must either cut back in size, or survive on stored resources, for example during the night. (speculative)

Induces flowering in pineapples

In food production, some plants are considered ethylene producers, while others are considered ethylene sensitive

Etiolate To blanch or whiten a plant or part of a plant by excluding sunlight. This is often done with endive, celery, and cauliflower.

Etiolation 1.The process of causing a plant to whiten by excluding sunlight.

2. Pale, sickly, excessive growth caused by insufficient light, incorrect feeding, or overcrowding.

Etiology The study of the causes of diseases.

Eubacteria One of two major groups of prokaryotes, including most common species of bacteria.

Euchromatin Decondensed, transcriptionally active interphase chromatin.

Eucyclic Having a regular alternation of parts; applied to flowers in which the petals, stamens, etc. are equal in number in each whorl and alternate with one another.

Eukaryote An organism containing a membrane-bound nucleus and other organelles, i.e. all higher plants, animals, fungi, and protists .

Eukaryotic cells Cells that have a nuclear envelope, cytoplasmic organelles, and a cytoskeleton.

European corn borer (ECB) Ostrinia nubilalis (Hübner) is a major insect pest in corn that costs farmers millions of dollars annually in control expenditures and yield loss. The European corn borer (ECB) larva feed on the leaves and 'borer' holes into the stalks where they tunnel.

Eusporangium (eusporangia) A meiosporangium that arises from several initial cells and forms a wall of more than one cell layer.

Eustele A stele in which the vascular bundles are arranged in discrete strands around a pith; typical of seed plants.

Euthrophication A condition in surface waters where high nutrient concentrations stimulate algal blooms which are sufficient to choke out some other forms of life. A more advanced state of nutrition pollution is a hypereutrophic condition where many clear water species cannot survive. Less contaminated states include oligiotrophic with clear waters and little biological activity and mesotrophic with an intermediate level of nutrient supply and biological activity.

Evaporation To convert from a liquid to a vapor

Evapotranspiration 1. The loss of water from the soil by evaporation and plant transpiration.

2. Water on the surface taken away by water, wind, air, and elements.

Event The insertion of a particular transgene into a specific location on a chromosome. Events are differentiated by two factors: 1. what transgene was inserted, and

2. where on the chromosome it inserted and how many gene copies inserted at that locus.

Everbloomer A plant that blooms continuously throughout the growing season.

Evergreen Keeping its leaves all year long, although losing some of the older leaves regularly throughout the year. Semi-evergreen plants only lose some leaves or lose older leaves only when new ones develop.

Everlastings This describes plants whose flowers maintain their color, brightness and form long after being gathered.

Exalate Lacking wings or wing-like appendages; not alate.

Exalbuminous Without endosperm; applied to seeds. Also endospermous, exendospermic.

Exanthem 1. A blotch on the surface of a leaf, etc.

2. A copper-deficiency disease of plants, prevalent in citrus and olive, and characterized by dieback and blotches on leaves and fruit.

Exchangeable A term used to describe ions in the absorbing complex of soil that can be exchanged with other ions.

Exchangeable sodium Sodium that is attached to the surface of soil particles and that can be exchanged with other positively charged ions in the soil solution.

Exclusion Control of disease by excluding the pathogen or infected plant material from crop production areas (e.g. by quarantines and embargoes).

Exfoliate To peel off in layers or flakes, as the bark of certain trees.

Exocarp The outer layer of the covering of a fruit when it consists of two dissimilar layers.

Exogen A plant in which the growth of the stem is in successive concentric layers, such as most trees.

Exogenous Applied externally; not from within the organism.

Exon A segment of a gene that contains a coding sequence.

Exonuclease An enzyme that hydrolyzes DNA molecules in either the 5′ to 3′ or 3′ to 5′ direction.

Exosporic Of gametophytes, forming outside the meiospore cell wall.

Exothermic Characterized by or formed with evolution of heat

Exotic Of foreign origin or character; introduced from a foreign country or a different region. An exotic plant is one not native to the place where it is growing such as Japanese Honeysuckle, which has naturalized in the northeastern states, or Eucalyptus trees, which have been naturalized on the West Coast.

Expansins A special class of wall-loosening proteins which cause cell wall expansion by loosening the hydrogen bonding between wall polysaccharides

Explants A piece of plant tissue cut for tissue culture.

Exportin A receptor protein that recognizes nuclear export signals and directs transport from the nucleus to the cytosol.

Expressed Genes for which there is current expression of the mRNA ('turned on'). For example, a gene may be expressed in a particular tissue or at a particular developmental stage but not at other times or in other tissues.

Expression vector A vector used to direct expression of a cloned DNA fragment in a host cell.

Exserted Extends beyond the throat.

Exsuccous Dry, without juice, sap, or moisture; withered.

Extensive farming System of farming, using few inputs of bought in feed and fertilizer, usually carried out over wide area on large upland farms in the UK.

Extracellular matrix (ECM) Region outside of metazoan cells which includes compounds attached to the plasma membrane, as well as dissolved substances attracted to the surface charge of the cells. The ECM functions both to keep animal cells adhered together, and well as buffering them from their environment.

Extracellular Outside a cell.

Extracellular polysaccharides (EPS) Sugar polymers that aid in movement, help prevent desiccation, and contribute to the slimy appearance of bacteria, fungi, and nematodes; and probably promote colonization of plant tissues and disease development in bacteria.

Extrachromosomal inheritance (also cytoplasmic or maternal inheritance) Inheritance of genes not located in the nucleus, i.e. those in mitochondria and chloroplasts.

Extrafloral Of a plant part that doesn't form part of a flower.

Extrorse Facing away from the direction of the growth (in reference to the anther of a flower).

Exudate Liquid excreted or discharged from diseased tissues, from roots and leaves, or by fungi.

Eye 1. The center of a flower when it is a different color from the petals.

2. An undeveloped bud on a tuber, as on a potato.

3. A cutting with a single bud.

Eyespot Light-sensitive organelle found in many groups of protists, and in some metazoans.

F

F.SP. (abbr. for forma specialis) A taxonomic group within a pathogenic species defined in terms of host range, i.e. members of different formae speciales infect different groups of plants.

F1 generation (first filial generation) The offspring produced from crossing two parental lines or individuals.

F2 generation (second filial generation) The offspring produced from self crossing or intercrossing individuals in the F1 generation.

Facilitated diffusion The transport of molecules across a membrane by carrier or channel proteins.

Facultative Capable of changing life-style, e.g. from saprophytic to parasitic or the reverse.

Facultative parasite Organism that is normally saprophytic but is capable of being parasitic.

Facultative saprophyte Organism that is normally parasitic but is capable of being saprophytic.

FAK (focal adhesion kinase) A nonreceptor protein-tyrosine kinase that plays a key role in integrin signaling.

Falcate Hooked; curved like a scythe or sickle; falciform.

Falciform Sickle-shaped; falcate.

Fallow The practice of leaving land either uncropped and weed-free, or with volunteer vegetation during at least one period when a crop would normally be grown; objective may be to control weeds, accumulate water, and/or available plant nutrients.

Fallow(ing) Resting land from deliberate cropping, not necessarily without cultivation or grazing, but without sowing

Falls The drooping or horizontal petals of irises.

Family This consists of genera (singular is genus, go see) that resemble one another in certain broad characteristics. The family usually consists of numerous genera, though some only consist of one. The family name usually ends with -aceae or -ae. For example, Compositae, which is the largest family of plants, consists of 950 genera and 20,000 species. Asters, Sunflowers, Mums, and Dahlias are all well-known members of that family and these genera contain numerous species.

Fan training A method of training a tree or vine on a trellis or wall in a way that causes the branches radiate from the trunk at regular intervals and at continually smaller angles, the lower branch on each side being horizontal.

Farinose Covered with mealy powder (i.e. leaves of certain plants

Fasciate Banded or compacted together; exhibiting fasciation. Also fasciated.

Fasciation A malformation in plants, in which a stem or branch becomes expanded into a flat, ribbon-like shape, as if several stems were joined together lengthwise.

Fascicle A close cluster or bundle, as of leaves, stems, or flowers. Sometimes limited in use to a compacted cyme.

Fastidious In reference to prokaryotic organisms, having special growth and nutritional requirements.

Fastigiate trees Trees grown for their narrow, columnar shape.

Fastigiate With erect branches and close together.

Fats See triacylglycerols.

Fatty acid In chemistry, especially biochemistry, a fatty acid is a carboxylic acid (or organic acid), often with a long aliphatic tail (long chains), either saturated or unsaturated. Long carboxylic acids as short as butyric acid (4 carbons) are considered to be fatty acids, while fatty acids derived from natural fats and oils may be assumed to have at least 8 carbon atoms, e.g. caprylic acid (octanoic acid). Most of the natural fatty acids have an even number of carbon atoms, because their biosynthesis involves acetate which has two carbon atoms.

Industrially, fatty acids are produced by the hydrolysis of the ester linkages in a fat or biological oil (both of which are triglycerides), with the removal of glycerol. See oleochemicals.

Saturated fatty acids

Saturated fatty acids do not contain any double bonds or other functional groups along the chain. The term "saturated" refers to hydrogen, in that all carbons (apart from the carboxylic acid [-COOH] group) contain as many hydrogens as possible. In other words, the omega (ω) end contains 3 hydrogens (CH_3-) and each carbon within the chain contains 2 hydrogens (-CH_2-).

Saturated fatty acids form straight chains and, as a result, can be packed together very tightly, allowing living organisms to store chemical energy very densely. The fatty tissues of animals contain large amounts of long-chain saturated fatty acids. In IUPAC nomenclature, fatty acids have an -oic acid suffix. In common nomenclature, the suffix is usually -ic.

Some saturated fatty acids are:

Butyric:
$CH_3(CH_2)_2COOH$

Lauric (dodecanoic acid):
$CH_3(CH_2)_{10}COOH$

Myristic (tetradecanoic acid):
$CH_3(CH_2)_{12}COOH$

Palmitic (hexadecanoic acid):
$CH_3(CH_2)_{14}COOH$

Stearic (octadecanoic acid):
$CH_3(CH_2)_{16}COOH$

Arachidic (eicosanoic acid):
$CH_3(CH_2)_{18}COOH$

Unsaturated fatty acids

Unsaturated fatty acids are of similar form, except that one or more alkenyl functional groups exist along the chain, with each alkene substituting a singly-bonded " -CH_2-CH_2-" part of the chain with a doubly-bonded "-CH=CH-" portion (that is, a carbon double bonded to another carbon).

The two next carbon atoms in the chain that are bound to either side of the double bond can occur in a *cis* or *trans* configuration.

Cis

A *cis* configuration means that the two carbons are on the same side of the double bond. The rigidity of the double bond freezes its conformation and, in the case of the *cis* isomer, causes the chain to bend and restricts the conformational freedom of the fatty acid. The more double bonds the chain has in the *cis* configuration, the less flexibility it has. When a chain has many *cis* bonds, it becomes quite curved in its most accessible conformations. For example, oleic acid, with one double bond, has a "kink" in it, while linoleic acid, with two double bonds, has a more pronounced bend. Alpha-linolenic acid, with three double bonds, favors a hooked shape. The effect of this is that in restricted environments, such as when fatty acids are part of a phospholipid in a lipid bilayer, or triglicerides in lipid droplets, cis bonds limit the ability of fatty acids to be closely packed and therefore could affect the melting temperature of the membrane or of the fat.

Trans

A *trans* configuration, by contrast, means that the next two carbon atoms are bound to *opposite* sides of the double bond. As a result, they don't cause the chain to bend much, and their shape is similar to straight saturated fatty acids.

In most naturally occurring unsaturated fatty acids, each double bond has 3*n* carbon atoms after it, for some n, and all are cis bonds. Most fatty acids in the *trans* configuration (trans fats) are not found in nature and are the result of human processing (eg, hydrogenation).

The differences in geometry between the various types of unsaturated fatty acids, as well as between saturated and unsaturated fatty acids, play an important role is biological processes, and in the construction of biological structures (such as cell membranes).

Examples of unsaturated fatty acids:

Alpha-linolenic acid:
$CH_3CH_2CH{=}CHCH_2CH{=}CHCH_2CH{=}CH(CH_2)_7COOH$

Docosahexaenoic acid

Eicosapentaenoic acid

Linoleic acid:
$CH_3(CH_2)_4CH{=}CHCH_2CH{=}CH(CH_2)_7COOH$

Arachidonic acid:
$CH_3(CH_2)_4CH{=}CHCH_2CH{=}CHCH_2CH{=}CHCH_2CH{=}CH(CH_2)_3COOH$

Oleic acid:
$CH_3(CH_2)_7CH{=}CH(CH_2)_7COOH$

Erucic acid:
$CH_3(CH_2)_7CH{=}CH(CH_2)_{11}COOH$

Alpha-linolenic, docosahexaenoic, and eicosapentaenoic acids are examples of omega-3 fatty acids. Linoleic acid and arachidonic acid are omega-6 fatty acids. Oleic and erucic acid are omega-9 fatty acids. Stearic and Oleic acid are both 18 C fatty acids. They differ only in that stearic acid is saturated with hydrogen, while oleic acid is an unsaturated fatty acid with two fewer hydrogens.

Essential fatty acids

The human body can produce all but two of the fatty acids it needs. These two, linoleic acid and alpha-linolenic acid, are widely distributed in plant and fish oils. Since they cannot be made in the body from other substrates and must be supplied in food, they are called essential fatty acids. In the body, essential fatty acids are primarily used to produce hormone-like substances that regulate a wide range of functions, including blood pressure, blood clotting, blood lipid levels, the immune response, and the inflammation response to injury infection.

Essential fatty acids are polyunsaturated fatty acids and are the parent compounds of the omega-6 and omega-3 fatty acid series, respectively. They are essential in the human diet because there is no synthetic mechanism for them. Humans can easily make saturated fatty acids or monounsaturated fatty acids with a double bond at the omega-9 position, but do not have the enzymes necessary to introduce a double bond at the omega-3 or omega-6 position.

The essential fatty acids are important in several human body systems, including the immune system and in blood pressure regulation, since they are used to make compounds such as prostaglandins. The brain has increased amounts of linolenic and alpha-linoleic acid derivatives. Changes in the levels and balance of these fatty acids due to a typical Western diet rich in omega-6 and poor in omega-9 fatty acids is alleged to be associated with depression and behavioral change, including violence. The actual connection, if any, is still under investigation. Further, changing to a more natural diet, or consumption of supplements to compensate for a dietary imbalance, has been shown to reduce violent behavior and increase attention span, but the mechanisms for the effect are still unclear. So far, at least three human studies have shown results that support this: two school studiesas well as a double blind study in a prison

Trans fatty acids

A trans fatty acid (commonly shortened to trans fat) is an unsaturated fatty acid molecule that contains a *trans* double bond between carbon atoms, which makes the molecule less 'kinked' in comparison to fatty acids with *cis* double bonds. These bonds are characteristically produced during industrial hydrogenation of plant oils. Research suggests that increasing amounts of trans fats are, for causal reasons not well understood, correlate with circulatory diseases such as atherosclerosis and coronary heart disease, than the same amount of non-trans fats.

Fatty acids can be bound or attached to other molecules, like triglycerides or phospholipids. When they are not attached to other molecules, they are known as "free" fatty acids.

The uncombined fatty acids or free fatty acids may come from the breakdown of a triglyceride into its components (fatty acids and glycerol).

Free fatty acids are an important source of fuel for many tissues since they can yield relatively large quantities of ATP. Many cell types can use either glucose or fatty acids for this purpose. However, heart and skeletal muscle prefer fatty acids. On the other hand, the brain cannot use fatty acids as a source of fuel, relying instead on glucose, or on ketone bodies produced by the liver from fatty acid metabolism during starvation, or periods of low carbohydrate intake.

Short chain carboxylic acids such as formic acid and acetic acid are miscible with water and dissociate to form reasonably strong acids (pK_a 3.77 and 4.76, respectively). Longer chain fatty acids do not show a great change in pK_a. Nonanoic acid, for example, has a pK_a of 4.96. However, as the chain length increases the solubility of the fatty acids in water decreases very rapidly, so that the longer chain fatty acids have very little effect on the pH of a solution. The significance of their pK_a values therefore only has relevance to the types of reactions in which they can take part.

Even those fatty acids that are insoluble in water will dissolve in warm ethanol, and can be titrated with sodium hydroxide solution using phenolphthalein as an indicator to a pale pink endpoint. This analysis is used to determine the free fatty acid content of fats, i.e. the proportion of the triglycerides that have been hydrolyzed.

Fatty acids react just like any other carboxylic acid, which means they can undergo esterification and acid-base reactions. Reduction of fatty acids yields fatty alcohols. Unsaturated fatty acids can additionally undergo addition reactions, most commonly hydrogenation, which is used to convert vegetable oils into margarine. With partial hydrogenation, unsaturated fatty acids can be isomerized from *cis* to *trans* configuration.

Fatty acids at room temperature undergo a chemical change known as auto-oxidation. The fatty acid breaks down into hydrocarbons, ketones, aldehydes, and smaller amounts of epoxides and alcohols. Heavy metals present at low levels in fats and oils promote auto-oxidation. Fats and oils often are treated with chelating agents such as citric acid.

Fauna Animal life.

Faveolate Honeycombed; alveolate. In botany, also favose.

Feed Food for livestock, usually consisting of grains such as corn, rye and barley.

Feedback inhibition A type of allosteric regulation in which the product of a metabolic pathway inhibits the activity of an enzyme involved in its synthesis.

Feeder root Fine root that absorbs water and dissolved nutrients.

Female A plant that produces fruit; the plant that bears the pistil and receives the pollen or fertilizing element of the male plant, or the analagous organ in cryptogams. Pertaining to the kind of plants that produce fruit; pistil bearing; pistillate; producing pistillate flowers.

Fermentation Oxidation of certain organic substances in the absence of molecular oxygen.

Fernery A place where ferns are growing, often in a greenhouse; a plantation of ferns.

Ferric iron An oxidized or high-volence form of iron.

Ferritin Storage form of iron in the liver and other tissues

Ferrous iron A reduced or low-volence form or iron.

Ferrous sulfate Most commonly used inorganic iron source; not effective as a soil-applied material because it quickly reverts to an unavailable form.

Ferruginous Rust-colored.

Fertigation Application of plant nutrients in irrigation water .

Fertile 1. Producing abundantly; capable of sustaining abundant growth, as fertile soil.

2. Capable of producing fruit; having a functional pistil, bearing functional pollen, etc.

3. Capable of fertilizing, as an anther with well-developed pollen.

Fertility The quality of soil that enables it to provide compounds in adequate amounts and in proper balance for the growth of specified plants.

Fertilization 1. The act of rendering land fruitful or productive; the application of fertilizer.

2. The process by which pollen reaches the ovules and assures the production of fruit; also, the analogous process in cryptogams.

Fertilizer A natural or man-made material added to soil to supply nutrients to plants.

Fertilizer analysis The percent composition of a fertilizer as determined in a laboratory and expressed as total N, available phosphoric acid (P2O5) equivalent, and water-soluble potash (K2O) equivalent.

Fertilizer grade An expression that indicates the percentage of plant nutrients in a fertilizer.

Fertilizer injector A device connected to the faucet end of a garden hose; it contains about a pint of liquid or water soluble fertilizer and mixes it at a predetermined rate with water flowing through the hose.

FI hybrid The first generation of offspring produced by crossing two parents of different genotypes.

Fiber Elongated plant cell with thick walls (in xylem and phloem), dead at maturity.

Fibril 1. A small fiber.

2. A root hair.

Fibrillose Furnished with or made up of fibrils.

Fibroblast A cell type found in connective tissue.

Fibronectin The principal adhesion protein of the extracellular matrix.

Fibrous Composed of profusely branched roots with many lateral rootlets but with no main or tap root development.

Fibrous roots A fine, densley branching root, which absorbs moisture and nutrients from the soil.

Field capacity Soil water content after the soil has been saturated and allowed to drain freely for 24-48 hours. Free drainage occurs because of the force of gravity pulling on the water. When water stops draining, we know that the remaining water is held in the soil with a force greater than that of gravity.

Field moisture The water that soil contains under field conditions.

Field water capacity (FC) The content of water, on a mass or volume basis, remaining in a soil 2 or 3 days after having been wetted with water and after free drainage is negligible .

Filament A structure consisting of many cells joined end-to-end, or a single, large multinucleate cell with the same thread-like appearance.

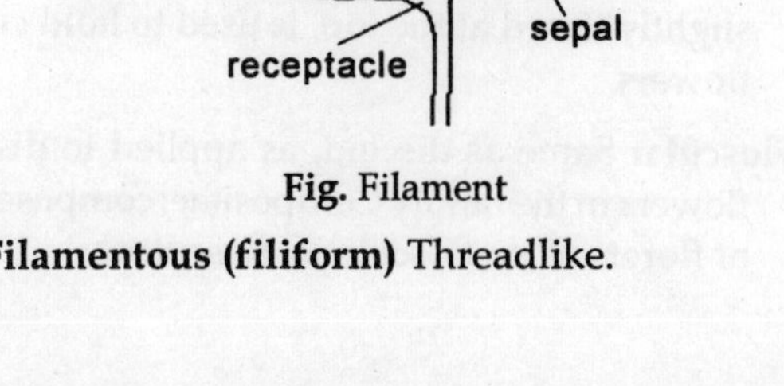

Fig. Filament

Filamentous (filiform) Threadlike.

Filiciform In the shape of a fern or fern frond.

Filiform Thread-like.

Film water Water held on the surface of the soil particles that doesn't drain away.

Filopodium A thin projection of the plasma membrane supported by actin bundles.

Fimble The male hemp plant or the fiber made from it. Also called fimble hemp.

Fimbria A fringe-like border; the peristome of a moss.

Fimbriate Fringed; bordered with hairs or with slender processes, as a fimbriate petal. Also fimbriated.

Fine sand Soil particles measuring from .10 to .25 millimeter.

Fine-textured soil A clayey soil containing 35% or more clay.

Firming A term used to indicate light-packing of soil around roots.

First bottom The normal flood plain of a stream.

Fish meal An organic fertilizer made of fish.

Fission Division of single-celled organisms, especially prokaryotes, in which mitosis does not occur. Also used to refer to mitosis in certain unicellular fungi.

Fitness The ability of an organism to survive and reproduce;

Fixation (of nitrogen) Conversion of molecular nitrogen (N_2) to ammonia and subsequently to organic nitrogen utilizable in biological processes.

Flabellate Fan-shaped. Also flabelliform.

Flaccid Body becomes soft and weak, lacking firmness

Flag Isolated, wilted or necrotic branch with dead leaves attached.

Flag leaf The uppermost leaf on the stem.

Flagellated Having flagella.

Flagellum n. Hair-like structure attached to a cell, used for locomotion in many protists and prokaryotes. The prokaryotic flagellum differs from the eukaryotic flagellum in that the prokaryotic flagellum is a solid unit composed primarily of the protein flagellin, while the eukaryotic flagellum is composed of several protein strands bound by a membrane, and does not contain flagellin. The eukaryotic flagellum is sometimes referred to as an undulipodium.

Flanking Flanking means sequences that border something. Flanking primers are located on either side of a DNA sequence of interest.

Flash grazing A relatively short grazing period usually involving high grazing pressure.

Flash tape A breeze-activated, metalized plastic tape that produces bursts of light in response to breezes; suspended over crops to scare away birds.

Flat A container for holding packs of plant starter cells. Also called seed flat.

Flavin adenine dinucleotide (FADH2) A coenzyme that functions as an electron carrier in oxidation/reduction reactions.

Flavonoids A group of chemical compounds found in all plants, most notably fruits, vegetables, nuts, seeds, and roots. Flavonoids are divided into 5 categories: flavonols, flavones, flavanones, flavan-3-ols, and anthocyanidins. Flavonoids are most commonly known for their antioxidant activity.

Fleck Minute, discolored spot in green tissue.

Flexuous Having turns or windings; capable of bending.

Floating row cover This is a lightweight fiber sheet, water and air penetrable, that is placed over a row of plants for protection from heat, cold or insects.

Floccose Covered with tufts of soft, woolly hairs.

Flocculate To clump together individual tiny soil particles.

Flocculose Finely floccose. Also flocculous.

Floccus A small tuft of woolly hairs.

Flood irrigation A shallow flood of water over level soil.

Flora Plant life in general, often in contrast to fauna. The plants native to or adapted to a particular region, habitat, etc. or belonging to a particular period.

Floral Containing, belonging to, or relating to a flower; pertaining to flowers in general; made of flowers.

Floral envelope The perianth of a flower.

Florescence Blooming; the unfolding of a flower; the state of being in bloom.

Floret The flower (including the lemma and palea) of cereals and grasses.

Florets Small or reduced flowers; often found in grasses and the Asteraceae family

Floriculture The cultivation of flowers or of flowering plants, usually commercially.

Floriculturist A person who specializes in growing flowers.

Floridean starch the storage polysaccharide of the Division Rhodophyta.

Floriferous Flower-bearing, usually used to indicate profuse flowering.

Floriform In the form of a flower.

Floriparous This pertains to plants that normally bear fruit but instead produce secondary flowers.

Florist One who sells flowers and ornamentals.

Florist's flower vase A tapered metal bucket, slightly flared at the top, is used to hold cut flowers.

Floscular Same as discoid, as applied to disk flowers in the family Compositae; composed of florets. Also flosculous, flosculose.

Flow cytometer An instrument that measures the fluoresence intensity of individual cells.

Flow diagram The diagrammatic representation of the structure of a system illustrating physical and information flows between compartments

Flower 1. The part of the plant that ordinarily contains the reproductive organs, which are usually surrounded by colorful petals and sepals. A flower containing only stamens, is male, or staminate; a flower containing only pistils, is female, or pistillate. When a flower contains both pistils and stamens it is called a bisexual, or perfect, flower; if it also has a corolla and calyx, it is called a complete flower.

2. Flower is also generally used to describe any blossom or inflorescence or any plant of which the blossom is the essential feature; a plant grown for its floral beauty.

Flower arrangement A combination of individual cut flowers to create an attractive decoration.

Flower break Break or stripe in flower color .

Flower heads A mass of small flowers or florets that together appear to form a single flower (e.g. Daisy); a form of inflorescence giving the effect of a single flower, consisting of a compact cluster of florets sessile upon the shortened summit of the axis.

Flower press A contraption in which flowers may be dried and preserved. A flower press consists of layers of cardboard and blotting paper sandwiched between two wooden boards, which are held together in the corners by adjustable screws.

Flower ring A wire loop supported by stakes to prop and contain a cluster of weak-stemmed flowers such as bachelor buttons or campanulas.

Flower stalk A peduncle or pedicel; the ordinarily leafless part of a stem or branch that bears a flower cluster or a single flower.

Flower support A stake, usually made of plastic coated wire, to prop or brace a single weak-stemmed plant such as lily or gladiolus. Also called loop stake.

Floweret A floret; sometimes refers to one of the segments of a cauliflower head.

Fluid mosaic model A model of membrane structure in which proteins are inserted in a fluid phospholipid bilayer.

Fluid-phase endocytosis The nonselective uptake of extracellular fluids during endocytosis.

Fluorescein A fluorescent dye.

Fluorescence in situ hybridization (FISH) A method used to localize genes on chromosomes using fluorescent probes.

Fluorescence Light emitted by an excited singlet molecule as part of the de-excitation process. Because a small amount of the energy in the original exciting photon is lost by molecular vibration before the emission of the fluorescence, the wavelength of fluorescence is greater (i.e. less energy) than the wavelength of the exciting photon.

Fluorescence microscopy Type of microscopy in which molecules are detected based on the emission of fluorescent light.

Fluorescent tube A type of light bulb that emits ultraviolet light. Fluorescent lamps are used to start and maintain plants indoors.

Fluorescent-light gardening The process of growing plants under artificial lights.

Fluorophore A molecule which fluoresces, give off light energy, when excited by an external light source.

Flux The rate of the flow of energy or chemical compounds from a source (a source gives) to a sink (a sink accepts).

Focal adhesion A site of attachment of cells to the extracellular matrix at which integrins are linked to bundles of actin filaments.

Focus Small area of diseased plants within a population.

Foliaceous Being or resembling a leaf. In botany, having the texture or form of a leaf, as a foliaceous sepal or bract.

Foliage The mass of growing plant leaves in their natural form and condition, or a cluster of the same; leafage.

Foliar Consisting of or pertaining to leaves.

Foliar feeding Feeding plants by spraying plant food on their leaves.

Foliate Leafy; covered with leaves or leaflets (i.e. foliate stalk).

Foliation The leafing of plants; the disposition of the beginning leaves within the bud.

Foliicolous Growing upon leaves; parasitic on leaves, as many fungi.

Foliolate Pertaining to, consisting of, or having leaflets. Used in combination, as bifoliolate, having two leaflets.

Foliole A leaflet; a small part resembling a leaf.

Foliose a form of lichen in which the thallus is thin but elevated above the substrate, and somewhat leaf-like in appearance.

Foliose Leafy or resembling a leaf. Also folious.

Follicle A dry, one-celled seed vessel consisting of a single carpel and dehiscent only by the ventral suture, as in the milkweed and larkspur.

Food chain Series of organisms, each eating or decomposing the preceding one. A simple example of a food chain is one in which a red-tailed hawk eats an eastern phoebe, which has eaten a great number of insects that lived in streams, which have all eaten stream algae.

Food web Complex network of many interconnected food chains and feeding relationships.

Footstalk The stalk, or petiole, of a leaf or the peduncle of a flower.

Forage Crops consumed by livestock

Foraging The search and consumption of forage.

Forcing The practice of bringing a plant into growth or flower (usually by artificial heat or controlling daylight) at a season earlier than its natural one.

Forcing house A hothouse used for forcing plants.

Forcing pit A pit of wood or masonry, sunk in the earth, for containing fermenting materials to produce bottom heat in forcing plants.

Forest decline Combination of biotic factors (e.g. insect pests, pathogens) and abiotic factors (e.g. drought, air pollution) that contributes to poor growth and death of populations of trees.

Form A botanical category ranking below a variety and differing only trivially from other related forms, such as in flower color.

Form genus A genus based on morphology and not on evolutionary relationships, such as used for imperfect fungi.

Forma specialis (abbr. f.sp.; pl. formae speciales) A taxonomic group within a pathogenic species defined in terms of host range, i.e. members of different formae speciales infect different groups of plants.

Fornicate Overarched fornices, as the throat of the corolla of the forget-me-not. Also forniciform.

Fornix A small, arching scale or appendage in the throat of the corolla of some plants, such as in forget-me-nots.

Fossil A remnant or trace of an organism of past geological ages that has been preserved in the Earths crust.

Fossil fuels Biological materials which have been subjected to long term geological effects. i.e coal, oil natural gas and peat

Fragipans The hard and brittle layers in soil that roots and water can't penetrate because it is so dense and compact.

Fragment banding pattern Bands made when many copies of the same DNA fragment move together through an electrophoresis gel. The pattern of bands reveals the the number of different size fragments in a sample.

Frame A garden frame is a bottomless box with a transparent top that is used to provide favorable conditions for the plant's growth and to cover them during harsh weather. The top is made of plastic or glass set in a wooden or metal frame. To allow for proper ventilation, the top is formed with one or two moveable sections called sash; the sashes are moved by sliding them up and down or, if they are hinged, by raising their fronts. Their are unheated and heated frames and they are both good for raising seedlings ahead of the outdoor season and for growing early crops of Lettuce, Cauliflower, Radishes, Corn, Snapdragons, and other vegetables and flowers.

Free Used to describe the condition of a substance within a mixture when it isn't chemically combined with other components of the mixture.

Free radical Any molecule with one unshared electron giving the molecule a partial negative charge; highly reactive and can cause permanent damage to cellular membranes.

Free range System of (mainly) poultry keeping in which hens are allowed to range over a relatively large area

Free water Unbound water; often used to describe a film of water on a plant surface.

Freeze fracture Method of electron microscopy in which specimens are frozen in liquid nitrogen and then fractured to split the lipid bilayer, revealing the interior faces of cell membranes.

Frequency A measure of the number of times per second (units of sec-1) that the electric field in a photon vibrates.

FRET Fluorescence Resonance Energy Transfer - a real time PCR DNA detection system, using flanking primers.

Friable Rich, loamy soil that is easily worked and loose; soil of a crumbly texture, able to be worked easily. Friable soil crumbles easily when worked with a spade, fork, rake, etc. The opposite is a tenacious, clayey soil. The particles of this soil cling together forming clods that are difficult to pulverize. Nonfriable soil can be enhanced by the following methods Mixing in organic matter such as manure, leaf mold, peat moss and compost; turning under cover crops or green manures such as Winter Rye, Ryegrass and Buckwheat and adding sand or gritty coal cinders; liming; and the use of synthetic soil conditioners.

Frond A leaf of a palm or fern.

Frondage A group of fronds.

Frondent Leafy.

Frondescence The period or state of coming into leaf; leafage; foliage.

Frondescent Coming into leaf.

Frondiferous Producing fronds or leaves.

Frondlet A small frond.

Frondose Bearing or resembling fronds.

Frost 1. That state or temperature of the air that occasions freezing of water.

2. The state or condition of being frozen; said of the surface of the ground when frost extends to a depth of 10 inches.

Frost control The act of minimizing frost damage to plants by covering them, watering them lightly, or placing them in a protected area.

Frost-free days The number of days in a growing season in any given area.

Fructescence The time fruit matures.

Fructiferous Bearing or producing fruit.

Fructification 1. The production of fruit by a plant; fruiting.

2. The ripened ovary and its appendages; the fruit of a plant.

Fructify To bear or produce fruit; to make fruitful.

Fruit 1. Any product of vegetable growth useful to humans or animals (i.e. grapes, figs, corn, cotton, flax, and all cultivated plants). There are many kinds of fruit and they differ greatly in character and degree of complexity.

2. The reproductive product of a plant; the seed of plants, or the part that contains the seeds.

3. An edible, juicy product of a plant, usually covering the seeds (i.e. apple, orange, a berry, a melon, etc.).

4. The matured ovary of a plant, consisting of the seeds and their pericarp (coating) and including whatever may be incorporated with it; also, the spores of cryptogams and the organs accessory to them. (verb) 1. To produce fruit; to come into bearing.

Fruit dot The sorus of ferns.

Fruit tree A tree grown for its fruit, or a tree whose main value consists in the fruit it bears (i.e. cherry tree, apple tree, pear tree, etc.).

Fruit trees, dwarf Describes genetic dwarf trees and trees grafted onto dwarf roots. For example, peaches that are genetic dwarfs produce full size fruit on trees that hardly ever grow more than 3 feet high. Fruit trees grafted onto dwarf rootstock, such as citrus or apples, produce normal fruit on trees that only grow up to 8 feet high. Many grafted fruit trees are also sold as semi-dwarf, growing to about 10 feet high. Dwarf fruit trees are great for growing in suburban yards, containers, and in areas with little space.

Fruit, compound A fruit that is made of several ovaries.

Fruiting body A specialized organ for producing spores, as in mosses and mushrooms.

Fruitlet 1. A small fruit.

2. A unit of a collective fruit.

Frumentaceous Having the character of or resembling wheat or other cereal.

Frustule The mineral "skeleton" of a diatom or other unicellular organism.

Frutescent Having the appearance or habit of a shrub; shrubby, as a frutescent stem.

Fruticose a form of lichen in which the thallus is branched and often cylindric, often resulting in a shrub-like appearance.

Fruticulose Resembling a small shrub.

Fucoxanthin a xanthophyll pigment characteristic of the Phaeophyta and also found in many Chrysophyta and Pyrrhophyta.

Fugacious Withering away early; ephemeral.

Fumigant (fumigate) A gas or volatile substance used to kill or inhibit the growth of microorganisms or other pests.

Functional Genomics An area of study that determines what trait or function a gene encodes.

Fungal Pertaining to, characteristic of, or consisting of a fungus or fungi. Also fungous.

Fungi (singular is fungus) A division of diverse plants that lack chlorophyll and do not photosynthesize. Included are mushrooms, molds, rusts, smuts, and yeasts. Some are edible, such as Boletus; many are plant diseases, such as powdery mildew.

Fungi Imperfecti (Deuteromycetes) Group of fungi without a sexual stage; also, the asexual stage of ascomycetes and basidiomycetes.

Fungicide Chemical or physical agent that kills or inhibits the growth of fungi.

Fungistasis Inhibition of fungal growth, sporulation, or spore germination but not death; used to describe the nonspecific phenomenon in natural soils where spore germination is inhibited and often overcome by rhizosphere nutrients.

Fungistat A chemical or physical agent that inhibits fungal growth, sporulation, or spore geminiation, but does not cause death.

Fungus A eukaryotic organism that is usually filamentous (forming a mycelium) and heterotrophic, has cell walls composed of chitin, and reproduces by sexual and/or asexual spores.

Funiculus (funiculi) The stalk that connects the ovule to the ovary wall.

Funnel form Having the shape of a funnel; applied to the corolla of a flower in which the tube enlarges gradually from below, as the morning glory; infundibuliform.

Furcate Forked.

Furfuraceous Coated with bran-like, flaky particles; scurfy.

Furrow 1. A trench in the earth, especially one make by a plow.

2. To cut a furrow in; make furrows in; plow.

Furrow weed A weed growing on plowed land.

Furrowed Having longitudinal channels, ridges, or grooves, as a furrowed stem.

Fuscous Brown; brown tinged with gray.

Fusiform Tapering toward both ends like a spindle, as the radish. Also fusate.

FWAG Farm and Wildlife Advisory Group.

G

G protein-coupled receptor A receptor characterized by seven membrane-spanning a helices. Ligand binding causes a conformational change that activates a G protein.

G0 A quiescent state in which cells remain metabolically active but do not proliferate.

G1 phase The phase of the cell cycle between the end of mitosis and the begining of DNA synthesis.

G2 phase The phase of the cell cycle between the end of S phase and the begining of mitosis.

Galbulus A spherical cone formed of thick scales with a narrow base, as in the Cypress, or berry-like with fleshy scales, as in the Juniper.

Galea A helmet-shaped part of a calyx or corolla, as the upper lip of a ringent or labiate corolla. The "helmet" formed by the dorsal sepal or fusion of dorsal sepal and petals in some Orchids.

Galeate Having a galea; helmet-shaped (i.e. the helmet-shaped upper sepal of Monkshood).

Galeiform Helmet-shaped.

Gall Galls are proliferations of cell tissue in plants and can be caused by various parasites, from fungi and bacteria, to insects and mites. Galls are often very organised structures and because of this, the cause of the gall can often be determined without the actual agent being identified. This applies particularly to some insect and mite galls.

Galls induced by insects are called insect galls. Insects that induce insect galls are called gall-inducing insects. Insect galls are usually induced by the chemicals injected by the larvae or the adults when they consume part of the plants. After the galls were formed, the larvae stay inside and develop until fully grown, at which time they leave. In order to form galls, the insects must seize the time when plant cell division occurs at a high speed, and when the plant tissues are still growing, normally spring. Although insect galls can be found on a variety of parts of the plant, such as the leaves, stalks, branches, buds, roots or even flowers, gall-inducing insects are fairly particular about which tissue of the plants or what kind of plants they make galls on. They would at most make galls on plants similar to those that they usually make galls on.

Galls are rich in resins and tannic acid and have been used in the manufacture of permanent inks and astringent ointments, in dyeing, and in tanning. A high-quality ink has long been made from the Aleppo gall, found on oaks in the Middle East; it is one of a number of galls resembling nuts and called gallnuts or nutgalls.

Gallery Insect tunnel in bark and wood.

Gametangium Cell containing gametes or nuclei that act as gametes.

Gamete A mature male or female reproductive cell (sperm or ovum) containing half of the total number of chromosomes in a cell (i.e., humans have 46 chromosomes per cell; gametes would have 23).

Gametophyte The generation or stage of a plant producing gametes, existing only as a rudimentary structure in seed plants and as distinguished from the sporophyte, or asexual form.

Gamopetalous Having the petals joined at the base.

Gamophyllous Having united leaves or leaf-like organs. Also monopetalous.

Gamosepalous Having the sepals united. Also monosepalous.

Gap junction A plasma membrane channel forming a direct cytoplasmic connection between adjacent cells.

Garden compost An organic matter such as well-decayed leaves, grass-clippings, and vegetable waste, added to the soil as an amendment to improve its texture and drainage and to enrich it with nutrients.

GATT General Agreement on Tariffs and Trade.

Gazebo An ornamental garden structure, resembling a free-standing porch, which may contain benches or statuary.

Geitonogamy The pollination of a pistil by pollen from another flower of the same plant.

Geminiflorous Having flowers in pairs.

Gemma (gemmae) Multicellular asexual reproductive structures of liverworts, consisting of disc-shaped masses of cells borne in a gemma cup.

Gemmaceous Relating to gemmae.

Gemmate Having gemmae; reproducing by buds.

Gemmation The act of budding; asexual reproduction in which a new organism originates locally within the body or on the surface of the parent. Also gemmulation.

Gemmiform Resembling a gemma.

Gemmoid Having the nature or form of a gemma.

Gemmule A small bud or gemma.

Gene The fundamental unit of heredity that carries genetic information from one generation to the next. A gene is an ordered sequence of nucleotides located on a particular position on a particular chromosome that encodes a specific functional protein.

Gene action The phenotypic response observed when the dosage of an allele changes. For example, in the allelic series AA, AB, and BB, the dosage of the B allele changes from 0 to 1 to

2. Gene action may be additive (if AA<AB<BB), dominant (if AB=BB), or overdominant (if AB>BB).

Gene amplification An increase in the number of copies of a gene resulting from the repeated replication of a region of DNA.

Gene cloning Finding and making copies of a specific gene. This is the first part of the genetic engineering process.

Gene Expression Conversion of the information encoded in a gene via transcription and translation, resulting in the production of a protein and the appearance of the phenotype determined by that gene.

Gene flow Movement of genes among and between species or populations.

Gene gun Also known as microprojectile bombardment and particle acceleration. The method used to transform cells using small gold or tungsten particles which are coated with DNA and literally shot into the cell.

Gene library A collection of cloned DNA molecules representing all or part of an individual's genome.

Gene mapping Determination of the relative positions of genes on a DNA molecule or chromosome and the distance between them.

Gene silencing The inactivation of a gene by an organism to prevent the gene from expressing.

Gene stacking Creating a plant with a desired combination of two or more unique genes. This can be done when the genes are first transferred into the cells by co-transformation, or during the breeding process by crossing two lines that each contain a different gene resulting in progeny with both genes.

Gene transfer The introduction of foreign DNA into a cell.

Gene-for-gene hypothesis The hypothesis that corresponding genes for resistance and virulence exist in the host and pathogen, respectively.

General homologous recombination The exchange of segments between DNA molecules that share extensive sequence homology.

General resistance (horizontal resistance, race non-specific resistance) Resistance that is effective against all biotypes of the pathogen.

General transcription factors Transcription factors that are part of the general transcription machinery.

Generative nucleus The nucleus of a pollen grain that divides to form the sperm cells.

Generic 1. Pertaining to a genus, or a kind or group of similar things; comprehending a number of like things, without specifying them.

2. Having the taxonomic rank or classificatory value of a genus (i.e. a generic name or description, generic characters or differences, generic identity). In the botanic classification of Genipa Americana, Genipa is the generic name of over 75 trees with similar characteristics, differing from other genera or groups of plants.

3. Distinctly characteristic; so marked as to constitute or denote a distinct kind.

Generic agreement/identity The agreement of objects that belong to the same genus.

Generic area The region to which the members of a genus (group) are limited in distribution. (The place in a generic area where the genus is most represented by species or individuals is known as its metropolis.)

Generic description/diagnosis A description or characterization of a genus in botany.

Generic diversity The differences between individuals of different genera.

Generic name The denomination that comprehends all the species, as of a group of animals, plants or fossils, which have generic characteristics in common.

Genesis The means of development and origin of soil.

Genetic Relating to heredity; referring to heritable characteristics.

Genetic code The sequence of nucleotides, read three at a time, that determines the order of amino acids in a protein.

Genetic engineering The science of modifying the genetic constitution of plants and animals directly

Genetic fingerprinting Determining the genetic makeup of an organism with the use of DNA markers.

Genetic Map A figure demonstrating the relative position of specific genes along a chromosome.

Genetic Marker A small region of DNA along a chromosome that can be used to indicate the location of a gene or other DNA sequence.

Genetic Modification The alteration to an organism's genome by any number of methods, including inserting, transferring, or deleting genes or other DNA sequences.

Genetically modified organism; (GMO) (transgenic) Organism possessing a gene from another species; used to describe the organisms that have been the subject of genetic engineering.

Geniculate Bent at an abrupt angle, like a knee. Also geniculated.

Genome All the genetic material in the haploid set of chromosomes for a particular organism.

Genomic imprinting The regulation of genes whose expression depends on whether they are maternally or paternally inherited, apparently controlled by DNA methylation.

Genomic library A collection of recombinant DNA clones that collectively contain the genome of an organism.

Genotype Genetic constitution of an individual or group; class or group of individuals sharing a specific genetic makeup .

Genus A taxonomic category that includes a group of closely related (structurally or phylogenetically) species; the genus or generic name is the first name in a Latin binomial.

Geocarp Fruit that ripens beneath the ground (e.g. peanut pod).

Geochemistry The chemistry of the composition of and changes to the solid matter of the earth.

Geophysics A field of geology using the principles of physics to explore the structure and physical characteristics of the Earth

Geothermal energy Energy contained in the earths heat as steam water and hot rock

Geotropic Pertaining to, or exhibiting, geotropism.

Geotropism 1. Tropism in which gravity is the orienting factor, up or down.

2. Tropism in which growth is toward the earth. Also geotropy.

Germ The term used by millers to denote the embryo of cereals.

Germ theory The theory that infectious or contagious diseases are caused by microorganisms (germs) .

Germ tube Hypha resulting from an outgrowth of the spore wall and cytoplasm after germination.

Germinate To cause to grow or sprout; to begin growing.

Germination 1. The act, process, or result of germinating; the development of a seed. The development of a seed into a plant. Oxygen, warmth and moisture are necessary for germination to occur. The moisture softens the seed, relaxes tissues in the embryo and dissolves nutrients. Oxygen is necessary for the chemical changes needed for growth. The temperature needed varies between the species. Some seeds, such as wheat seeds, will germinate on melting ice; others need temperatures higher than 60° F. When germination takes place, a root is produced that grows down into the soil; at the same time, a stem and leaves are growing upward.

2. The similar development of a plant from a spore in cryptogams, such as ferns.

3. The early period of growth in a bud, as a bulb or a rhizome.

4. The protrusion and growth of the pollen tube from the pollen grain.

Germinative Pertaining to germination; capable of germination.

Germline cells Cells that give rise to gametes, the reproductive cells of an organism, (such as an egg or sperm), that have only one set of chromosomes. When the two unite, they form a living embryo.

Germplasm Bearer of heredity material, often loosely applied to cultivars and breeding lines.

Geyser A type of hot spring seen on land that episodically erupts jets of hot water and steam. It results from ground water being heated by coming in contact with hot rock deep in the Earth's crust, and then rising to the surface.

Giant-cell Enlarged, multinucleate cell formed in roots by repeated nuclear division without cell division induced by secretions of certain sedentary plant-parasitic nematodes .

Gibberellin A plant growth substance (phytohormone) involved in promotion of stem elongation, mobilisation of food reserves in seeds and other processes. Its absence results in the dwarfism of some plant varieties. Chemically all known gibberellins are gibberelliç acids, a family of diterpenoids acids that are synthesized by the terpenoid pathway in plastids and then modified in the endoplasmic reticulum and cytosol until they reach their biologically-active form.

Much of our knowledge of the biosynthesis and molecular mechanisms of gibberellins comes from research on their role in triggering á-amylase release by the aleurone layer in seed germination. Gibberellin was first isolated in 1926 by Japanese scientists. It was derived from the *Gibberella* fungus lamapitus.

Synthesized in the embryo and germinating seeds

Synthesized in the roots

Increased in production in the dark when sugar cannot be manufactured, and decreased in production in the light

Synthesized in apical meristems and young leaves, as well as root tips and seeds.

Transported in non-polar, bidirectional manner, producing general responses

Released in all cells (more particularly root and mature cells) when they do not have enough sugar and oxygen to support both themselves and any dependent cell (speculative)

Released in response to root, environmental, pest, and disease stress

Directly induced by high levels of CK (speculative)

Effects

Stimulates shoot and cell elongation

Delays senescence of leaves

Inhibits root growth

Inhibits adventitious root growth

Produces seed germination

Antagonist promotes root growth and GA reverses this

Promotes root initiation in low concentration in pea cuttings

Stimulates bolting and flowering in biennials

Regulates production of hydrolytic enzymes for digesting starches

Inhibits CK bud growth on calluses

Inhibits bud formation

Inhibits leaf formation

Breaking of dormancy

Gibbous Swollen, usually at the base (as in the corolla).

Gibbs free energy (G) The thermodynamic function that combines the effects of enthalpy and entropy to predict the energetically favorable direction of a chemical reaction.

Gill Thin, radial membrane producing basidiospores in the cap of certain basidiomycetous fungi (e.g. mushrooms).

Girdle To circle and cut through a stem or the bark and outer few rings of wood, disrupting the phloem and xylem.

Glabrescent Becoming glabrous.

Glabrous Hairless; smooth. Also glabrate.

Gladiate Sword-shaped; ensiform.

Gland Any secreting organs of plants or of gland-like bodies.

Glandular 1. Resembling or pertaining to a gland.

2. Bearing glands.

3. Derived from glands. Also glandulous.

Glans A nut seated in or enclosed by an involucre.

Glaucescent Slightly glaucous.

Glaucous 1. Blue-green, blue-gray, gray, or pale yellow-green color.

2. Having a waxy or powdery, bluish-white or bluish-gray bloom that can be rubbed off (i.e. glaucous plums or grapes).

Gley soil A layer of soil that has turned gray because of lack of oxygen and excessive water.

Global Warming An increase in the average temperature of the earth's atmosphere, especially a sustained increase sufficient to cause climatic change. Most scientists believe that a rise in carbon dioxide levels (caused by automobile, power plant, and other emissions) will lead to further global warming.

Globose Spherical, or nearly so.

Globular stage The developmental stage of embryogenesis that the embryonic tissues organized as a ball.

Glochidiate Barbed at the tip. Also glochideous.

Glochids Barbed bristly hairs or spines, often very fine and easily detached, on an Opuntia cactus.

Glomerate Compactly clustered.

Glomerulate Arranged in small, compact clusters. Also glomerulose.

Glomerule A compact flower cluster that forms a head, as in the flowering Dogwood, Cornus florida.

Gluconeogenesis The synthesis of glucose.

Glucoside A substance that, on decomposition, yields glucose and certain other compounds; some glucosides are defense compounds produced by plants, e.g. cyanogenic or phenolic glucosides.

Glumaceous Glume-like; having glumes; chaffy.

Glume A chaffy bract as in the Gramineae (grass family) and related plants; especially an empty bract at the base of a grass spikelet.

Glumiferous Having glumes.

Glutathione transferase An enzyme that catalyzes the binding of xenobiotics to glutathione.

Glutinous Sticky.

Glycerol phospholipids Phospholipids consisting of two fatty acids bound to a glycerol molecule.

Glycocalyx A carbohydrate coat covering the cell surface.

Glycogen A polymer of glucose residues that is the principal storage form of carbohydrates in animals.

Glycolipid A lipid consisting of two hydrocarbon chains linked to a polar head group containing carbohydrates.

Glycolysis The anaerobic breakdown of glucose.

Glycoprotein A protein linked to oligosaccharides.

Glycosaminoglycan (GAG) A gel-forming polysaccharide of the extracellular matrix.

Glycosidic bond The bond formed between sugar residues in oligosaccharides or polysaccharides.

Glycosylation The addition of carbohydrates to proteins.

Glycosylphosphatidylinositol (GPI) anchor Glycolipids containing phosphatidylinositol that anchor proteins to the external face of the plasma membrane.

Glyoxylate cycle The glyoxylate cycle is a metabolic pathway occurring in plants and several microorganisms. The glyoxylate cycle allows these organisms to use fats for the synthesis of carbohydrates, a task which vertebrates, including humans, cannot perform.

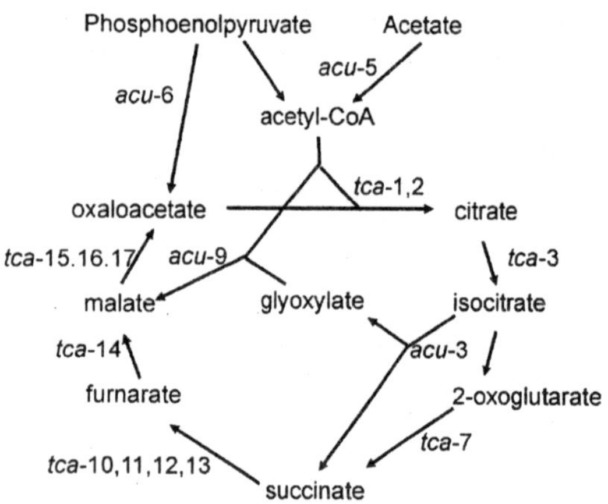

Fig. Glyoxylate cycle

When fatty acids are consumed by vertebrates they are degraded to many copies of a small 2-carbon compound: acetate. Acetate binds to coenzyme A and enters the citric acid cycle, where it is fully oxidized to carbon dioxide, which is released into the environment. This pathway allows the cell to obtain energy from fat.

Many cells in the body require direct supply of glucose, either for synthesis of polysaccharides or for energy production when glycogen is depleted. Gluconeogenesis is a pathway which enables the production of glucose from smaller molecules such as pyruvate, lactate and glycerol.

Pyruvate is the initial compound in gluconeogenesis. It is converted to oxaloacetate, which is in turn converted to phosphoenolpyruvate (PEP). Seven further reactions bring about the production of glucose. Oxaloacetate is also the initial and at the same time end product of the citric acid cycle. Since acetate groups can enter the citric acid cycle and eventually be converted to oxaloacetate (which can continue to produce glucose in gluconeogenesis), it may seem that the production of glucose from fatty acids is possible.

However, this does not happen in vertebrates. Acetate groups which enter the citric acid cycle are, as mentioned above, fully oxidized to form carbon dioxide. The acetate is therefore lost and cannot be converted to oxaloacetate, and later on to glucose.

Several organisms, however, found a solution to this, in the form of the glyoxylate cycle. The two initial stages of this cycle are identical to those of the citric acid cycle: acetate —>citrate. The next step, however, is different: isocitrate is cleaved into succinate and glyoxylate (the latter gives the cycle its name). Succinate can be channeled directly into the citric acid cycle and eventually form oxaloacetate. Glyoxylate condenses with acetyl-CoA, yielding malate. Both malate and oxaloacetate can be converted into phosphoenolpyruvate and gluconeogenesis can be initiated. The net result of the glyoxylate cycle is therefore the production of glucose from fatty acids.

In plants the glyoxylate cycle occurs in special peroxisomes which are called glyoxysomes. Vertebrates cannot perfrom the cycle because they lack its two key enzymes: isocitrate lyase and malate synthase.

Glyphosate A herbicide which inhibits the production of amino acids.

Golgi apparatus In cell biology, the Golgi apparatus (also called a Golgi body, Golgi complex, or dictyosome) is an organelle found in most eukaryotic cells, including those of plants, animals, and fungi. The name comes from Italian anatomist Camillo Golgi, who identified it in 1898. The primary function of the Golgi apparatus is to process proteins targeted to the plasma membrane, lysosomes or endosomes, and those that will be formed from the cell, and sort them within vesicles. Thus, it functions as a central delivery system for the cell. It is part of the endomembrane system.

Most of the transport vesicles that leave the endoplasmic reticulum (ER), specifically *smooth* ER, are transported to the Golgi apparatus, where they are modified, sorted, and shipped towards their final destination. The Golgi apparatus is present in most eukaryotic cells, but tends to be more prominent where there are many substances, such as proteins, being secreted. For example, plasma B cells, the antibody-secreting cells of the immune system, have prominent Golgi complexes.

The Golgi apparatus is considered more or less the "post office" of the cell. It handles all incoming lipids, proteins, etc., and controls their export, as well.

The transport vesicles from the Endoplasmic Reticulum (ER) fuse with the *cis* face of the Golgi apparatus (to the cisternae) and empty their protein content into the Golgi lumen. The proteins are then transported through the medial region toward the *trans* face and are modified on their way. Possible modifications include glycosylation and phosphorylation. The proteins are also labeled with a sequence of molecules according to their final destination. For example, the Golgi apparatus adds a mannose-6-phosphate label to proteins destined for lysosomes.

The transport mechanism itself is not yet clear; it could happen by cisternae progression (the movement of the apparatus itself, building new cisternae at the *cis* face and destroying them at the *trans* face) or by *vesicular transport* (small vesicles transport the proteins from one cisterna to the next, while the cisternae remain unchanged). It is also proposed that the cisternae are interconnected, and the transport of cargo molecules within the Golgi is due to diffusion, while the localisation of Golgi-resident proteins is achieved by an unknown mechanism.

Once the proteins reach the *trans* face, they are embedded into coated transport vesicles and brought to their final destinations. The form of the vesicle is determined by the type of protein and the label it acquired.

An example of the Golgi complex's functioning is the modification of glycoproteins (used in cell membranes). Vesicles from the ER contain simplified glycosylated proteins. In the Golgi complex, carbohydrates are attached and removed from these glycoproteins, creating a diversity of carbohydrate structures on the proteins. After they have been secreted in to the cell the vesicles fuse to the cell membrane and release their contents.

Along with protein modification, Golgi apparatus is involved in the transport of lipids around the cell, as well, creating lysosomes, organelles involved in digestion.

Golgi body (also called the golgi apparatus or golgi complex) A flattened, layered, sac-like organelle that looks like a stack of pancakes and is located near the nucleus. The golgi body packages proteins and carbohydrates into membrane-bound vesicles for "export" from the cell.

Gonad Sexual structure in animals that produces gametes.

G-protein A family of cell signaling proteins regulated by guanine nucleotide binding.

Graft 1. Method of propagation by which an artificial union is made between different parts of individual plants. The stem or shoot that is inserted into a rooted plant is called the scion; the plant or part of a plant into which the scion is inserted is called the stock or understock. There are many different methods of grafting including flat grafting, split grafting, and side grafting, among others.

2. The point where a scion is inserted in the stock.

Graft hybrid A hybrid produced by grafting a branch of one species (scion) onto another (rootstock), in which the two grow together to form a single plant or tree.

Graft transmission Transmission of a pathogen from one host plant to another through fusion of living tissue from the diseased host with living tissue of a healthy host.

Grafting tape Tape backed with biodegradable cloth; used in budding and grafting operations and in banding tree wounds. Also called nurseryman's tape.

Grafting thread Fine waxed string used in budding and grafting operations.

Grafting wax A waxy substance used to hold and seal tree or rose grafts.

Grain 1. The unhusked or threshed seed of one of the cereal plants wheat, oats, rye, barley, maize, or millet; a corn.

2. Corn in general; the gathered seeds of cereal plants in mass.

Gram negative Bacteria staining red or pink in the Gram staining procedure after treatment with Gram's stain.

Gram positive Bacteria staining violet or purple in the Gram staining procedure after treatment with Gram's stain.

Gram stain Procedure used for identification of bacteria in which crystal violet stain, Gram's iodine, ethyl alcohol and safranin stain are applied in succession to cells of the bacteria.

Gramineous Grass-like; belonging to the family Gramineae. Also graminaceous, gramineal.

Graminicide Herbicides that target the grass family (Graminae); refers to the lipid synthesis inhibitor herbicides that or only effective on grasses.

Grandular hairs A hair with a minute bead of clear liquid at its apex.

Graniform Having the form of a seed or grain.

Granular Containing, bearing, resembling, or composed of grains or granules. Covered with small grains or granules. Also granulose, granulous.

Granular fertilizer A fertilizer made up to particles 0.1 inch in diameter.

Granulate 1. To form into granules; make rough on the surface.

2. To become formed into grains; become granular; consisting of or resembling grains or granules. Also granulated.

Granulation 1. The process of granulating.

2. One of the little elevations in a granulated surface.

Granule A little grain; a fine particle.

Granum A stack of thylakoid disks within the chloroplast is called a granum.

Grassed waterway A natural or constructed waterway, usually broad and shallow, covered with grasses, used to conduct surface water from or through cropland.

Grassland Land on which the vegetation is dominated by grasses, grasslike plants, and/ or forbs. Herbaceous vegetation provides at least 80% of the canopy cover.

Gravid Containing an egg or eggs; capable of depositing eggs.

Gravitational water The water, in large pores of the soil, which drains away under the force of gravity.

Grazing A method of feeding by herbivores

Grazing capacity The optimal stocking rate that will achieve a target level of animal performance (e.g., average daily gain) or other specific objective, while preventing deterioration of the ecosystem.

Grazing distribution Dispersion of animals during grazing within a management unit or area

Grazing efficiency The percentage of the total standing crop by weight that is ingested .

Grazing pressure Animal demand per unit weight of forage dry matter either at an instant in time (AU/T) or over a period of time (AUM/T).

Great soil group Any one of several broad groups of soil with fundamental characteristics in common.

Green fluorescent protein (GFP) A protein from jellyfish that is commonly used as a marker for fluorescence microscopy.

Green manure An organic crop material, turned under the soil, to be used as a fertilizer to add nitrogen to the soil. The green plants decay in the soil and provide humus and make stiff soil more porous and friable. A quick-growing crop is sown and dug into the ground before it reaches maturity (when it will provide the greatest bulk without being to tall to turn under).

Green Seed A canola production problem in which an early frost inhibits the degradation of the green pigment chlorophyll. The presence of chlorophyll in the oil lowers the yield and grade of oil obtained from the seed.

Greenhouse A building of glass or plastic that is used to grow and protect tender plants.

Greenhouse effect Process by which significant changes in the chemistry of Earths atmosphere may enhance the natural process that warms our planet and elevates temperatures. If the effect is intensified and Earths average temperatures change, a number of plant and animal species could be threatened with extinction. Certain gaseous components of the atmosphere, called greenhouse gases, transmit the visible portion of solar radiation but absorb specific spectral bands of thermal radiation emitted by the Earth. The theory is that terrain absorbs radiation, heats up, and emits longer wavelength thermal radiation that is prevented from escaping into space by the blanket of carbon dioxide and other greenhouse gases in the atmosphere. As a result, the climate warms. Because atmospheric and oceanic circulations play a central role in the climate of the Earth, improving our knowledge about their interaction becomes essential.

The group of about twenty gases responsible for the greenhouse effect through their ability to absorb long wave terrestrial radiation. They are all minor gases and together make up less than 1 per cent of the total volume of the atmosphere. Carbon dioxide (CO_2) is the most abundant, but methane (CH_4), nitrous oxide (N_2O), the chlorofluorocarbons (CFCs) and tropospheric ozone (O_3) also make significant contributions to the greenhouse effect. Water vapour also exhibits greenhouse properties, but has received less attention than the others. Since the beginning of the twentieth century, rising levels of these gases in the atmosphere, associated with increasing fossil fuel use, industrial development and agricultural activity, have brought about an enhancement of the greenhouse effect. The amount of carbon dioxide recycled during photosynthesis is also being reduced because of worldwide deforestation, which allows the extra gas to remain in the atmosphere. The net result has been a gradual global warming, projected to continue as long as the volumes of greenhouse gases in the atmosphere continue to rise. Plans to slow the warming include a reduction in greenhouse gas emissions and a slowing down of deforestation, but the results are, as yet, insignificant.

Greenhouse Effect Warming that results when solar radiation is trapped by the atmosphere, caused by the presence in the atmosphere of gases such as carbon dioxide, water vapor, and methane that allow incoming sunlight to pass through but absorb heat radiated back from the earth's surface. Many human activities cause levels of these gases to rise, resulting in an increase in the Earth's temperature.

Gross Domestic Product (GDP) The total final output of goods and services produced within a country in a year by residents and non residents regardless of allocation of domestic and foreign claims

Gross National Product (GNP) The total value of goods and services produced by an economy over a particular period of time. GNP is made up of consumer and government purchases, private, domestic and foreign investments in the country and the total value of exports

Ground cover Refers to low-growing or trailing plants that are used to carpet the ground, as underneath trees or as an overplanting for bulbs.

Ground meristem The primary embryonic tissues that surround the plant stem and branches.

Ground water 1. Water that fills all unblocked pores of underlying material below the water table.

2. The upper limit of saturation.

Grow light A broad-spectrum florescent light used to grow plants indoors.

Growing point The tip of a shoot from which new growth emerges.

Growing tray A tray having compartments like an ice-cube tray, used for starting seeds. Also called a cell tray.

Growth factors Polypeptides that control animal cell growth and differentiation.

Growth form The form of a plant; the habit in which a plant grows; also called growth habit or just habit. Shrub and vine are growth forms (i.e. a shrubby plant or a climbing plant - a vine).

Growth promoters Substances given to livestock to increase growth rates

Growth regulator (hormone) A chemical substance produced in one part of an organism and transported in minute quantities to induce a growth response in another part, e.g. in plants, auxins, cytokinins, and gibberellins.

Growth regulators Mode of action of herbicides that mimic plant hormones causing unregulated growth.

Growth ring The layer of wood developed in a tree's growing season, usually one year.

Grumrose Made of clustered, coarse grains or granules. Also grumous.

Guanine A purine that base-pairs with cytosine.

Guano A strong, organic fertilizer composed of bird droppings.

Guanylyl cyclase An enzyme that catalyzes the formation of cyclic GMP from GTP.

Guard cell One of a pair of cells that surround a stoma, an opening in the epidermis. By changing shape, the guard cells open and close the stoma.

Guard cells Specialized epidermal cells that form the boundaries of stomata, the pores in plant leaf surfaces

Gully A channel resulting from erosion and caused by the concentrated but intermittent flow of water usually during and immediately following heavy rains. Deep enough (usually >0.5 m) to interfere with, and not to be obliterated by, normal tillage operations.

Gully erosion The erosion process whereby water accumulates and often recurs in narrow channels and, over short periods, removes the soil from this narrow area to considerable depths, often defined for agricultural land in terms of channels too deep to easily ameliorate with ordinary farm tillage equipment, typically ranging from 0.5m to as much as 25 to 30 m.

Gum Gelatinous, sugary aggregate that is synthesized and exuded by plant tissues .

Gumbo soil 1. Adobe soil.

2. An alkaline soil found in the West and Southwest of the U.S. that has enough nutrients to grow plants but is too compact for most gardening purposes.

Gummosis Pathologic condition characterized by excessive formation of gums; the products of cell degeneration.

Guttate Spotted, as by drops. Also guttated.

Guttation Exudation of watery, sticky liquid from hydathodes, especially along leaf margins .

Guttiferous Producing gum or resinous substances.

Gymnocarpous Having a naked fruit, especially in lichens.

Gymnogynous Having a naked ovary.

Gymnosperm A plant belonging to the Gymnospermae, characterized by naked seeds, as in the conifers (plants producing cones).

Gymnospermous Having naked seeds. Also gymnogenous, gymnospermal, gymnospermic.

Gymnospore A naked spore.

Gymnosporous Having naked spores.

Gynandrous Having the stamens adnate to and apparently borne upon the pistils, as in all Orchids.

Gynantherous Having the stamens converted into pistils.

Gyno Pistil or ovary.

Gynobase A prolongation or enlargement of the flower receptacle that bears the gynoecium.

Gynodioecious Having perfect and female flowers growing upon separate plants.

Gynoecium (gynoecia) The megasporangiate part of the flower, consisting of one or more pistils made up of carpels.

Gynomonoecious Having both perfect and female flowers upon the same plant.

Gynophore An elongation of the receptacle of a flower, bearing the gynoecium at its apex, as in some Capparidaceae.

Gynostegium A covering of the gynoecium.

Gynostemium The column of an orchid, formed by the union of the androecium and gynoecium.

Gyroma 1. The annulus of a fern.

2. The shield of some lichens.

H

H+-ATPase A membrane protein located in the plasma membrane and tonoplast, used to carry out primary active transport of H+ out of the cytosol by coupling directly to the energy released during ATP hydrolysis

Habit The characteristic growth or general appearance of a plant. The "habit of growth" may be low-growing, tall, erect, spreading, trailing, bushy, and so on.

Habitat A place with a particular kind of environment suitable for the growth of an organism.

Half-hardy annuals Plants that will survive some frost, but not a long freeze.

Half-life The time it takes for a compound to degrade to half of its original concentration.

Halophilous Preferring or habitually growing in salty soil as maritime plants.

Halophytic Having characteristics of salt tolerance.

Hamate Curved at the tip, like a hook.

Handpicking A method of controlling pests by literally removing the pest from the plant by hand. This is effective with larger pests, such as tomato hornworms and snails.

Hanging basket A container hung from a brace, tree, etc., with a rope or cord, which is used to grow ornamental plants.

Hanging plant rotator A device that rotates a hanging plant a certain amount (usually 90°) to ensure even lighting.

Haploid Describes a nucleus or cell with a single set of unpaired chromosomes The haploid number is designated as n. Reproductive cells, formed as a result of meiosis, are haploid. Fusion of two such cells restores the normal (diploid) number.

Haptonema Peg-like structure unique to the Prymnesiophyta; its function is not known.

Hard dough A description the grain at maximum dry weight. The grain is hard to puncture and has started to dry off.

Hardening off The gradual process of acclimating plants started indoors to outside conditions. Setting them in a sheltered location outdoors for increasing lengths of time over a period of days is said to be hardening off the plants. If the plants are just set outside without getting used to the different atmospheres, they will most likely suffer.

Hardiness Ability to withstand stress.

Hardiness zones Areas classified by climate and temperature range for the purpose of determining which plants can be grown where. The U.S. is divided into eleven zones.

One is the coldest and eleven is the warmest. Plants living in zones one through four are considered vary hardy; those in zones 5 - 7 are hardy; those in zones 8 - 9 are fairly hardy; those in zone 10 are said to be very tender, which means most will not survive freezing; zone 11 is frost-free since it consists of subtropical or tropical regions. The zones are classified by their average low temperature. Zone 1 has an average low temperature of -50° F. Zone 2 = -50° to -40°; zone 3 = -40° to -30°; zone 4 = -30° to -20°; zone 5 - -20° to -10°; zone 6 = -10° to 0°; zone 7 = 0° to 10°; zone 8 = 10° to 20°; zone 9 = 20° to 30°; zone 10 = 30° to 40°; zone 11 = no lower that 40°.

Hardpan A hard or cement-like soil or soil layer.

Hardwood The wood from trees with encapsulated seeds (i.e. nuts or fruits), as distinguished from a conifer or cone-bearing tree, which has naked seeds. Oaks, Walnuts and Maples are all hardwoods. Pines and firs are softwoods.

Hardy Used to describe a plant that is able to survive the cold temperatures of a given location without protection. When a range is given, hardiness also describes a plant's ability to tolerate the hottest temperatures normally experienced. (For example, a Palm tree isn't hardy in Alaska, because it is a tropical plant requiring a warm, humid atmosphere..)

Harrow 1. A rake-like tool that is drug over plowed soil to prepare it for planting.

2. To draw a harrow over soil; to break or tear with a harrow, as to harrow ground.

Harvest ripe The grain has dried off and is ready to be harvested.

Hastate Having a triangular shape like a arrowhead.

Haulm The stems and leaves of a crop left after harvest.

Haustorium A specialized branch of a fungal hypha formed inside a living cell of the host plant in order to obtain nutrients.

Head 1. The top of a plant consisting of a compact mass of leaves, such as a head of lettuce.

2. A short, dense bunch of flowers.

3. To form a head.

4. To cut back the shoots of plants in order to induce lateral growth.

Headage payment Payment made to farmers based on the number of ewes or suckler cows kept on the farm

Headwater The source and upper part of a stream.

Heart stage A developmental stage in embryogenesis that the young embryo looks like a heart. At this stage, most all organs in a young plant have been formed.

Heartwood Central cylinder of nonfunctional xylem in a woody stem .

Heat mat A low-voltage heater used to warm soil during seed germination, seedling development, and root growth of cuttings.

Heat shock A method of cell transformation that opens small temporary holes in the plant cell wall by alternating temperatures between hot and cold. Foreign DNA mixed with these cells is able to pass into the cell through these holes.

Heating cable An electric device used to heat soil in a greenhouse or cold frame.

Heating system A unit consisting of an adjustable temperature controller and a heat mat; used to maintain soil temperature during germination and seedling development or during root growth of cuttings.

Heat-shock proteins A highly conserved group of chaperone proteins expressed in cells exposed to elevated temperatures or other forms of environmental stress.

Heaving The pushing of plants out of the soil because of alternate freezing and thawing action.

Heavy soil Fine-textured, clay soil.

Hectare Unit of area measurement, equivalent to ~ 2.5 acres.

Hedge 1. A barrier or fence formed by bushes or small trees growing together; a closely planted row of shrubbery.

2. To enclose or fence with a hedge; to separate by a hedge.

Hedge binding Something that is used to bind together the bushes composing a hedge.

Heel When a cutting it taken with a "heel" of wood, it means a piece of the old branch or shoot is detached from the old branch or shoot along with the cutting.

Heeling in Temporarily covering the base of a plant, especially a dormant plant, with soil for a short time.

Hefted sheep Hill or mountain sheep that are grazed in the same area that they were born, which is unfenced, they do not stray. A heft is a flock of such sheep sold with the farm

Helicase An enzyme that catalyzes the unwinding of DNA.

Heliophilous Fond of the sun; attracted by or becoming most active in sunshine.

Heliophobic Shade-loving.

Heliotropic Turning toward the sun, as leaves or flowers.

Heliotropism The tendency of plant parts, such as foliage, stems, or flowers, to position themselves according to the direction in which the sun hits them.

Helobious Living in swamps or marshes.

Hemibiotroph A parasite that initially forms an association with living cells of the host, much like a biotroph, and then in the later stages of infection it becomes necrotrophic, actively killing host cells.

Hemicellulose A polysaccharide that crosslinks cellulose microfibrils in plant cell walls.

Hemidesmosome A region of contact between cells and the extracellular matrix at which keratin filaments are attached to integrin.

Hemiparasite Parasitic flowering plant that contains chlorophyll when mature (photosynthetic) and obtains water with dissolved nutrients by connecting to the host xylem via the haustorium, e.g. true mistletoe (Phoradendron spp. or Viscum spp.) or witchweed (Striga spp.).

Hemizygous An organism that has a gene copy in only one chromosome in a chromosome pair (i.e., during the transformation process, the transgene will insert into only one chromosome of a pair).

Henequen A strong fiber, similar to sisal, which is produced from the leaves of a Central American Agave and used mainly for making twine. Also spelled heniquen.

Henry's law The moles of a particular gas that can be dissolved into a solvent depends on the partial pressure of that gas in the atmosphere

Herb 1. A plant grown for its medicinal or flavoring properties or for its scented foliage.

2. Also refers to plants without persistent, woody stems. This botanical use of the word "herb" is used to distinguish between trees and shrubs.

Herbaceous A characteristic that distinguishes a herb from a woody; no persistant, woody tissue above ground

Herbage 1. Leafy vegetation.

2. The herbaceous parts of a plant.

Herbal Pertaining to or consisting of herbs.

Herbalism The knowledge of herbs.

Herbalist 1. Someone who is skilled in the knowledge of plants or makes collections of them.

2. A dealer in medicinal plants; a person who treats diseases with botanical remedies only.

Herbarium A collection of dried plants systematically arranged. The plants are attached to sheets of paper, either by small gummed strips of paper or by gluing one side of the specimen.

Herbicide A chemical substance used to destroy or inhibit the growth of plants.

Herbicide family A group of herbicides that have similar chemical structure and kill plants by the same mechanism of action.

Herbicide resistance Is the inherited ability of a plant to survive and reproduce following exposure to a dose of herbicide normally lethal to the wild type. In a plant, resistance may be naturally occurring or induced by such techniques as genetic engineering or selection of variants produced by tissue culture or mutagenesis.

Herbicide tolerance The inherent ability of a species to survive and reproduce after herbicide treatment. This implies that there was no selection or genetic manipulation to make the plant tolerant; it is naturally tolerant.

Herbivores Animals that feed on plant material

Herbivorous Feeding on plants. For example, animals such as moose and snowshoe hares are herbivorous.

Herbivory The consumption of herbaceous vegetation.

Herborist An herbalist.

Herborization An excursion for studying or collecting plants.

Hercogamous Incapable of self-pollination.

Hercogamy A condition in which structural obstacles within a flower make self-pollination impossible.

Heritability The proportion of phenotypic variation that is due to genetics, as opposed to environment or genotype x environment interaction.

Hermaphrodite Bisexual, bearing both male and female organs in the same flower; having fully developed stamens and pistils in the same flower.

Hesperidium A fruit with a leathery rind (i.e. an orange).

Heterauxesis The irregular growth of a plant due to unequal rates of growth in its parts.

Heterochromatin Condensed, transcriptionally inactive chromatin.

Heterocyst Specialized cells found in the cyanobacteria in which nitrogen fixation occurs.

Heterodromous Having the genetic spiral of the branches different from that of the main stem, affecting the way in which leaves are arranged around a stem.

Heterodromy A difference in direction of the genetic spiral in the branch and the main stem.

Heteroecious Pertaining to a rust fungus requiring two unrelated host plants for completion of its life cycle .

Heterogamous Bearing two kinds of flowers in which the sexual parts are different.

Heterogamy The condition of being heterogamous.

Heterogeneous The state of being mixed in composition. For example, heterogeneous forests are comprised of many different species of trees.

Heterokaryon Cell with genetically different nuclei.

Heterologous genes A gene transferred from another organism which expresses the same trait as an endogenous gene.

Heteromerous In reference to flowers, having the parts of adjoining whorls different in number.

Heteromorphic Having irregular, abnormal, or unusual structure, as the leaves of a young plant.

Heteromorphic alternation of generations gametophyte and sporophyte generations differ in their vegetative form.

Heteromorphism The condition of being heteromorphic.

Heterophylly The presence of more than one type of leaf on the same plant, such as floating or aerial leaves and submersed leaves on the same aquatic plant.

Heterosis (hybrid vigor) Increased vigor of hybrid offspring when compared to either parent.

Heterosporous Having more than one kind of asexually produced spores (applied to ferns).

Heterostyled Having the styles of the flower of two or more different forms.

Heterotactic Characterized by heterotaxis.

Heterotaxis Abnormal arrangement of parts or organs as compared to a normal type.

Heterothallic Producing male and female or (+) and (-) gametes on different gametophytes, often of dissimilar appearance.

Heterothallism Condition in which sexual reproduction can occur only in the presence of genetically different mycelia .

Heterotroph An organism that obtains its carbon and energy from organic compounds by feeding on the tissues of autotrophs (producers) or of other heterotrophs (consumers).

Heterotrophic Receiving nutrition by the ingestion or absorption of high-energy organic compounds produced by other organisms.

Heterozygote An organism having two different alleles at one or more locations.

Heterozygous An organism that has two different alleles at one or more locations on a chromosome.

Hexane A organic molecule that forms a nonpolar solvent [CH3(CH2)4CH3]

Hexapetaloid Of a flower, having six petals. Also hexapetalous.

Hibernaculum The part of a plant that is dormant in the winter, as a bulb or bud.

Hierarchy A structural relationship in which each unit consists of two or more sub units, the latter been similarly divided

High-energy bonds Chemical bonds that release a large amount of free energy when they are hydrolyzed.

Hilar Belonging to the hilum, or scar, on a seed produced by the attachment to the pod, as with beans.

Hill 1. A small raised mound or raised row with a cluster of cultivated plants, such as potatoes.

2. A group of vining plants or others planted in a cluster instead of a row, such as a hill of squash or pumpkin, even if the soil isn't mounded above the level of the garden.

Hilum The scar on a seed marking its point of attachment, as with beans or peas.

Hirsute Having coarse, rigid hairs.

Hispid Having stiff, bristly hairs or tiny spines.

Hispidity The state of being hispid.

Hispidulous Somewhat hispid.

Histones Proteins that package DNA in eukaryotic chromosomes.

HLCA Hill Livestock Compensatory Allowance.

HMG-14 and **HMG-17** Nonhistone chromosomal proteins associated with decondensed transcriptionally active chromatin.

Hoary Covered with a whitish or grayish-white down.

Holdfast The portion of a filamentous or parenchymatous alga that attaches it to the substrate.

Holliday junction The central intermediate in recombination, consisting of a crossed-strand structure formed by homologous base pairing between strands of two DNA moleucles.

Holomorph The whole fungus in all its states .

Holoparasite Parasitic flowering plant that lacks chlorophyll (nonphotosynthetic) and must rely totally on the contents of the xylem and the phloem of the host plant, e.g. broomrape (Orobanche spp.).

Homeobox Conserved DNA sequences of 180 base pairs that encode homeodomains.

Homeodomain A type of DNA binding domain found in transcription factors that regulate gene expression during embryonic development.

Homeothermy Refers to body temperature that is constant and largely independent of the temperature of the animal's surroundings.

Homogamous 1. Of a plant, having all the flowers alike.

2. Of a flower, having the stamens and pistils maturing at the same period.

Homogamy The state of being homogamous.

Homogenous The state of being constant in composition. For example, a patch of forest is homogenous if it contains only one tree species. A soil sample is homogenized by mixing it thoroughly.

Homogony The state of being homogonous.

Homologous Having the same function, structure, position or value

Homologous recombination Recombination between segments of DNA with homologous nucleotide sequences.

Homology Similarity of appearance as a result of common descent.

Homomallous Referring to the leaves of mosses, curving to one side.

Homopetalous Having all the petals formed alike.

Homoplasy False homology.

Homosporous Producing only a single form of meiospore.

Homostyled Having all styles one length.

Homothallic Producing male and female gametes or isogametes on the same gametophyte.

Homothallism Condition in which sexual reproduction occurs with a single thallus; self-fertile .

Homozygosity Having like alleles at corresponding loci on homologous chromosomes. An organism can be homozygous at on, several, or all loci.

Homozygote An organism having identical alleles at one or more locations.

Homozygous An organism that has two identical alleles at one or more locations on a chromosome. Inbred lines are homozygous at many of their gene loci.

Honeydew 1. A sweet substance found on the leaves of trees and other plants in small drops, like dew. It is the sugary substance secreted from the leaves of plants, occurring most often in hot weather. It usually is seen as small, glistening drops, but it may drip abundantly from the leaves, when it has been called manna. This especially happens with Fraxinus Ornus (the Manna Ash) and Carduus arctiodes.

2. A sweetish secretion exuded by many insects, such as aphids, mealybugs, scale and whiteflies. The secretion attracts ants, which harvest the honeydew and protect the pests from natural predators. It also develops a blackish mold, known as sooty mold. Plants

that are heavily infested, especially trees, drip honeydew, staining anything underneath, such as vehicles. The best control is to eliminate ants (banding with sticky materials is effective) from the plant and then control the source pest. Insecticidal soaps will help clean the foliage.

Honeydewed Covered with honeydew, usually referring to buds and foliage.

Hook climber A plant that climbs by the aid of hooks or prickles, such as certain types of roses.

Horizons, soil The layers of soil, usually three, which make up the soil profile. Soil horizons differ in color, texture, structure, and organic matter content. The A Horizon is the upper surface or soil and usually has the highest organic matter content; the B Horizon is the subsoil; and the C Horizon is the parent material. A given soil may have one or all three horizons.

Horizontal resistance (general resistance, race non-specific resistance) Resistance which is effective against all biotypes of the pathogen (see durable resistance, specific resistance, vertical resistance).

Hormone Any of certain chemical substances that are produced by living cells in one part of the plant and initiate cellular activity (such as growth or flowering) in a distant part of the plant.

Hormones A chemical substance produced in a part of a plant that initiates and controls growth and flowering in another part of the plant

Horticultural oil A refined oil sprayed on plants to control a variety of pests. It kills by smothering the pests and their eggs or by disrupting membrane functions. It may also prevent some fungal diseases. Highly refined types are called summer, superior, or verdant oils and can be used during the growing season as long as temperatures are below 90° F. Heavier types are called dormant oils and are used on dormant, deciduous plants.

Horticulture The cultivation of fruit, flowers, vegatables and shrubs, also used to describe the commercial production of such crops on general farms

Horticulturist A person who specializes in horticulture.

Hortus siccus A collection of specimens of plants carefully dried and preserved for botanical purposes; and herbarium.

Hose guides Devices, anchored by stakes, used to route a garden hose above or around plants. The notched, or open-loop-top, type is used to route a hose above or around plants. The spool type is used to guide a hose around corners.

Hose in hose A flower that appears to have a double corolla, one within another, as in certain Azaleas.

Host A plant that supports a parasite that lives in or upon it.

Host plant Living plant attacked by or harboring a parasite or pathogen and from which the invader obtains part or all of its nourishment.

Host Range The range of plants on which an organism feeds, particularly a parasite ; the range of plants in which a pathogen is capable of causing disease.

Hotbed A bed of soil heated by fermenting substances and covered with glass to keep out cold air; intended for raising early plants or for protecting tender exotics.

Hotcap An individual, conical plant cover used to protect against frost, sun, birds, and insects.

Hothouse A structure kept heated for the growth of tender, exotic plants or subtropical plants, or for the production of native fruits, flowers, etc., out of season. The temperature of the hothouse is kept between the temperatures of the greenhouse and the stove-house, or orchid-house.

Houseplant Any plant that is suitable for growing in the conditions of a house.

HPLC/MS High performance liquid chromatograph coupled to a mass spectrometer

Hull 1. The outer covering of a seed, nut, or fruit.

2. To remove the hull of.

Humid acids 1. The term sometimes interchangeable for humus.

2. The alkali-soluble end products of the decomposition or organic matter in mineral soils.

Humidity meter An instrument used to measure the amount of moisture in house or greenhouse air.

Humming line A ribbon suspended over crops that emits am ultrasonic sound to discourage birds.

Humus An organic material in soil which is a product of plant and animal remains which have been decomposed and then synthesized into something new.

Husk 1. The outer covering of the fruit of a plant (i.e. corn husks).

2. To remove the husk of.

Hyaline Transparent or nearly so; translucent; often used in the sense of colorless

Hybrid An individual produced by crossing two parents of different genotypes.

Hybrid vigor (heterosis) Increased vigor of hybrid offspring when compared to either parent.

Hybridizable Capable of producing hybrid offspring by crossing with another variety, species, or genus.

Hybridization The practice of crossbreeding between two species or wild types; this term is used to indicate crossbreeding generally, whether between species or varieties.

Hybridize To produce or cause to produce a hybrid.

Hybridizer A person who hybridizes plants.

Hybridoma A cell produced by the fusion of an antibody-producing cell and a lymphoma (cancer) cell for production of monoclonal antibodies.

Hybrids Individuals produced by crossing two parents of different genotypes.

Hydathode A specialized epidermal leaf structure with one or more openings through which water is discharged from the leaf interior to its surface.

Hydric Wet; applied to features of plants or places where they grow.

Hydrogen An element, found in the soil, that plants need in order to grow.

Hydrogen bonding An electrostatic attraction between a hydrogen atom in one polar molecule of water and a small electronegative atom (oxygen) in another molecule of water

Hydrogen fluoride (HF) A common pollutant from volcanoes and several industries sucli as brick factories, aluminum smelters, and phosphate fertilizer factories.

Hydrogen sulfide A colorless, poisonous gas with an odor of rotten eggs.

Hydrograph A chart or graph that depicts changes in water quantity over time.

Hydroid A water-conducting cell of a moss.

Hydrological cycle The cycle of water evaporated from water surfaces and plants and precipitated as rain

Hydrology The scientific study of the properties, distribution, and effects of water on the earth's surface, in the soil and underlying rocks, and in the atmosphere.

Hydrolysis (hydrolyze) A chemical reaction in which some reactant combines with water and splits the water into hydrogen and hydroxyl ions.

Hydrolyze A chemical reaction in which some reactant combines with water and splits the water into hydrogen and hydroxyl ions.

Hydrophilic Having chemical properties relating to water (water-like), compounds are polar or ionizable and can be soluble in water; lipophobic.

Hydrophily The flowers are produced under the surface of the water and that is where the transfer of pollen takes place. Fairly rare occurrence.

Hydrophobic 'water fearing'; compounds that are relatively insoluble in water and other polar solvents and soluble in lipids and other nonpolar solvents

Hydrophyte A plant that grows in and is adapted to an aquatic or very wet habitat.

Hydroponics A method of growing plants in water rather than soil. The plant's roots are suspended in a nutrient solution.

Hydroseeding (or hydromulching) Is a process of spray grass-seeding with a mixture or blend of cellulose fiber mulch, grass seed and water. This forms a mat or blanket on the soil to enhance seed germination, vigorous growth, and good rooting system.

Hydrostatic pressure The positive pressure built up when water is compressed

Hydrothermal Pertaining to hot water.

Hydrothermal deposit A mineral deposit formed from precipitation of minerals from fluids.

Hydrothermal plume A cloud of hot, mineral-rich water that flows out of a hydrothermal vent and disperses into the ocean, usually several hundred meters above the seafloor vent site. Rock particles and minerals in the plume water often make the plume look smoky, like a cloud of ash over an erupting volcano.

Hydrothermal system A system of circulation of fluid caused by a heat source (often a magma chamber or hot rock) in which water flows into the crust in one area (the recharge zone) and out of the crust in another area (the discharge zone).

Hydrous Containing water.

Hydroxyapatite A member of the apatite groups or mineral rich in the hydroxyl groups.

Hygrophyte A plant growing in wet places.

Hymenium (hymenia) The layer of asci in an ascocarp.

Hypanthial Pertaining to or resembling a hypanthium.

Hypanthium The part of certain flowers usually formed by the fusion of the lower portions of their petals, sepals, and stamens.

Hyperhydricity A network formed by hydroxyl connection.

Hyperparasite Organism that is parasitic upon another parasite.

Hyperplasia Abnormal increase in the number of cells, often resulting in the formation of galls or tumors.

Hypersensitive The state of being extremely or excessively sensitive. It often refers to an extremmusation of disease in susceptible plants; a type of avirulence gene.

Hypersensitive response (HR) Rapid and localized cell death at the site of infection in resistant interactions between plants and pathogens.

Hypertrophy Abnormal increase in the size of cells in a tissue or organ, often resulting in the formation of galls or tumors.

Hypha A single tubular thread-like filament of a fungal mycelium. The hypha is the basic structural unit of a fungus.

Hyphal sheath (mantle) Dense hyphal mass of ectomycorrhizal fungus enclosing the short feeder roots of plants.

Hyphomycetes Imperfect fungi that produce conidia freely or on a sporodochium or synnema .

Hyphopodium Stalked, thick-walled, lobed cells that stick to plant surfaces; sometimes used to describe the infection structures produced by ectotrophic hyphae of certain root-infecting fungi such as Gaeumannomyces (take-all pathogen).

Hypochile The basal part of the labellum (of an Orchid) in some groups, such as Stanhopea.

Hypocotyl Portion of the stem below the cotyledons and above the root .

Hypocotyledonary Located below the seed leaves.

Hypodermis The cell layer beneath the epidermis of the pericarp

Hypogenous Growing on the under surface, especially pertaining to a fungus on the underside of a leaf.

Hypogynous Having the sepals, stamens and petals arising from the portion of a flower situated below the ovary.

Hypogyny The state of being hypogynous.

Hypophyllous Situated on the underside of a leaf.

Hypophyseal region The tissue of a developing embryo that is derived from the topmost cell of the suspensor and is later incorporated into the root apex of the embryo, giving rise to part of the root cap, its initial cells and the ground meristem initial cells.

Hypovirulence Reduced ability to cause disease.

Hypoxia Oxygen deficiency of aquatic life due to pollution of water by nutrients and sediments.

Hypsophyll A leaf located underneath spore-bearing leaves.

Hypsophyllary Pertaining to the hypsophyll.

I

IACS Integrated Administration and Control System.

Ice out The springtime date of winter lake ice cover breakup; the date after which a lake is no longer covered by ice.

ICM Integrated Crop Management

Icosahedral (icosahedron) Having 20 faces, as a polyhedral-shaped virus particle.

Icterus Jaundice.

IFM Integrated Farm Management.

Igneous rock A rock formed by the cooling and crystallization of molten magma.

Illuvial horizon An accumulation of material in the B horizon (subsoil) that moved into it from the A horizon (topsoil).

Imbricate Overlapping in regular order, as the scales on a snake.

IMC (Isolated Microspore Culture) Systems Offer an unlimited supply of embryogenic cells and developing embryos, which facilitate studies of key developmental processes revelant to seed formation and metabolism.

Imidazolinones A family of herbicides that are similar in chemical structure (imidazolinone ring) and have the same mechanism of action Immarginate Having no margin.

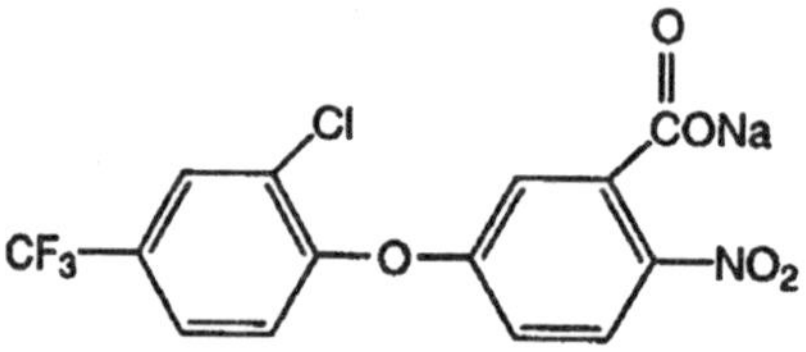

sodium 5-[2-chloro-4-(trifluoro-methyl)-phenoxy]-2-nitrobenzoate

Fig. Imidazolinones

Immature soil A soil that lacks clear individual horizons.

Immiscible Two solvents that are not able to dissolve into each other when mixed .

Immobilization The conversion of an element from the inorganic to the organic form in microbial tissues, so that the element is not readily available to other organisms.

Immortelle Any one of the flowers called everlasting. These flowers feel papery and keep their natural color and appearance after drying, therefore they are widely used for wreaths and other floral arrangements.

Immune Cannot be infected by a given pathogen.

Immunoassay A detection method based on antibodies specifically selected to react with the substance to be detected (the antigen).

Immunoblotting A method that uses antibodies to detect proteins separated by SDS-polyacrylamide gel electrophoresis.

Immunoglobulin See antibody.

Immunoprecipitation The use of antibodies to isolate proteins.

Immunosuppressant Chemical or activity that suppresses the natural immune responses in animals, including humans.

Imparipinnate Having pairs of leaflets opposite each other along a central stem, with a single leaf at the end.

Fig. Imparipinnate

Imperfect Diclinous; having the pistils and stamens in separate flowers. Flowers having only male or female parts.

Imperfect flower A flower lacking stamens or pistils.

Imperfect fungi (Fungi Imperfecti, deuteromycetes) Group of fungi without a sexual stage; also, the asexual stage of ascomycetes and basidiomycetes.

Impervious soil A soil through which water, air, or roots penetrate very slowly or not at all.

Importin A receptor protein that recognizes nuclear localization signals and directs nuclear import.

Impressed Sunken (such as veins).

Impression fossil The cast or mold of the surface of an organism in usually fine-grained sedimentary rocks.

In planta In a plant.

In situ hybridization The use of radioactive or fluorescent probes to detect RNA or DNA sequences in cell extracts, chromosomes, or intact cells.

In situ In its original place or environment.

In vitro In glass, on artificial media, or in an artificial environment; outside the host.

In vitro mutagenesis The introduction of mutations into cloned DNA in vitro.

In vitro translation Protein synthesis in a cell-free extract.

In vivo In a living cell or organism.

Inactivate Process of making a herbicide nonherbicidal by degradation or bonding to another molecule.

Inarable Not arable; not capable of being plowed or tilled.

Inarch To graft by uniting a scion to the stock, without separating the scion from its parent tree.

Inarching This is a method of grafting new plant growth onto a stronger root system by establishing young plants (usually, one that is in a pot) near an existing tree and at the point where they meet, removing matching areas of bark. The two cut surfaces are then fitted together and bound with soft tying material until they grow together and can be gradually separated with the new branches attached to the older rootstock.

Inbred An organism that has been self-fertilized for several generations until it is homozygous at all important gene loci, 'genetically pure'.

Inbreed To breed from individuals of the same parentage or closely related.

Inbreeding Deliberate self-pollination of a plant by transferring pollen from one of its flowers to the stigma of the same flower or to another flower of the same plant. Repeating this process for several generations ensures that the line will breed true.

Incanous Whitish or grayish.

Incised Sharply and usually deeply and irregularly cut, such as the edge of a leaf.

Included Not sticking out or projecting, as when pistils are included in the corolla.

Inclusion body Structure developed within a plant cell as a result of infection by a virus, often useful in identifying the virus.

Incomplete dominance Expression of a heterozygous phenotype which is different from, and often intermediate to, that of either homozygous parent.

Incomplete flower A flower lacking any of the four whorls or parts (sepals, petals, stamens, or pistils).

Incubation period The time between penetration of a host by a pathogen and the first appearance of disease symptoms; the time during which microorganisms inoculated onto a medium are allowed to grow.

Incubous Overlapping, like shingles.

Incumbent Lying on another organ, such as an incumbent anther lying against a filament.

Incurved Applied to petals and florets that curve inward to form a compact, rounded shaped.

Indeciduous Not deciduous; not losing its leaves annually; evergreen, as trees.

Indehiscent Fruits that don't burst open; not popping or splitting open spontaneously when mature, as a capsule or an anther.

Indeterminate A plant part that continues to grow while flowering, such as indeterminate tomatoes, which continue to grow until they die, or a flower spike that continues to grow longer while it's lower blossoms open.

Indeterminate growth Continuation of ontogeny when an organism reaches its final adult form; characteristic of plants.

Indexing Testing of a plant for infection, often by mechanical transmission or by grafting tissue from it to an indicator plant.

Indicator plant Plant that reacts to a pathogen or an environmental factor with specific symptoms, used to detect or identify the pathogen or determine the effects of the environmental factor.

Indigenous Native; not exotic.

Indirect penetration Penetration of plant tissues by a pathogen through natural openings (e.g. stomata) or wounds.

Indoor gardening The art of growing plants indoors using natural and/or artificial light.

Induced Produced in response to a stimulus .

Induced systemic resistance (ISR) Reduced disease symptoms on a portion of a plant distant from the area where the inducing agent is active, caused by the triggering of active plant defenses against a variety of pathogens; used to describe increased resistance in plants induced by certain rhizobacteria .

Inducing medium The tissue culture media for helping the start of organ or embryonic cell development.

Indumentum Thick hairy covering.

Induplicate Rolled inward, as the leaves in a bud.

Indusium A small flap covering the sori, or fruit dots, in ferns; a collection of hairs united so as to form a sort of cup and enclosing the stigma of a flower.

Infect Process in which an organism enters, invades, or penetrates and establishes a parasitic relationship with a host plant.

Infection court Site in or on a host plant where infection can occur.

Infection cushion Organized mass of hyphae formed on the surface of a plant from which numerous infective hyphae develop.

Infection focus Initial site of infection, generally with reference to a population of plants.

Infection peg (penetration peg) The specialized, narrow, hyphal strand on the underside of an appressorium that penetrates host cells.

Infection period The time required for infection to occur under conducive environmental conditions, usually hours of leaf wetness and temperature.

Infectious Pertaining to disease, capable of spreading from plant to plant.

Infective Referring to an organism able to attack a host and cause infection; referring to a vector carrying or containing a pathogen and able to transfer it to a host plant.

Inferior Below; beneath; lower.

Inferior ovary A basal, ovule-bearing part of a pistil borne below the point of attachment of the perianth and stamens.

Infest (infestation) To attack as a pest (used especially of insects and nematodes); to contaminate, as with microorganisms; to be present in large numbers .

Infiltration Penetration of water into and through soil pore spaces. How much water penetrates depends on several factors, including the rate of precipitation, soil moisture condition and soil texture.

Infiltration rate The rate of water controlled by surface conditions that enters the soil.

Inflated Enlarged or swelled out, as an inflated perianth.

Inflorescence The arrangement of flowers on a stem; inflorescence types include corymb, cyme, head, panicle, solitary, raceme, spike thyrse, and umbel

Influx Movement from the cell wall region into the cylasm; into of the cell.

Infrafoliar Below the leaves, as infrafoliar flowers.

Infrared mulch A plastic sheet that is placed around plants to warm the soil by infrared radiation and to inhibit weed growth by blocking most visible radiation.

Infraspecific Referring to categories of classification below that of species, such as subspecies.

Infrastructure The availability of roads, power supplies, education and health facilities, for example which all industries share and for which they do not directly pay for

Infructescence The fruiting stage of a plant.

Infundibuliform Having the form of a funnel.

Ingraft To insert, as a scion of one tree or plant into another, for propagation. Also spelled engraft.

Initial inoculum (primary inoculum) Inoculum, usually from an overwintering source, that initiates disease in the field, as opposed to inoculum that spreads disease during the season.

Injury Damage caused by transitory interaction with an agent such as an insect, chemical, or unfavorable environmental condition.

Inner bark Interior, living portion of the bark; secondary phloem.

Inoculate (inoculation) To place inoculum in an infection court; to insert a pathogen into healthy tissue.

Inoculum A pathogen or its parts which can cause infection when transferred to a favourable location ; the population of microorganisms introduced in an inoculation.

Inoculum density A measure of the number of propagules of a pathogenic organism per unit area or volume .

Inorganic 1. Not natural; artificial.

2. Composed of other than animal or plant matter; mineral.

Inorganic compounds Compounds not defined as organic. For example, water is inorganic.

Inorganic fertilizer A chemical fertilizer that quickly releases nitrogen into the soil.

Inorganic nitrogen The element nitrogen in combination with other mineral elements.

Inositol 1,4,5-trisphosphate (IP3) A second messenger, formed from the hydrolysis of PIP2, that signals the release of calcium ions from the endoplasmic reticulum.

Insect Member of the class Hexapoda (phylum Arthropoda) possessing three sets of limbs attached to a central body segment.

Insecticidal soap A low-toxicity pesticide made from the salts of fatty acids. It kills soft-bodied insects, like aphids, spider mites, thrips, and whiteflies, by disrupting cell membranes. (You can make your own by mixing 2 tablespoons dishwashing detergent in 1 gallon of water.) It may burn foliage in hot weather. It is sometimes combined with other materials, such as citrus oil or sulfur, to increase efficiency. Complete coverage of the plant is necessary for it to work.

Insecticide An agent that kills insects; specifically, that which kills insect pests of plants and humans.

Insectivorous (plants) Plants that are specially adapted for catching insects. Most, but not all, are grown for interest rather than beauty. Once their prey is captured, the plants secrete a fluid, which breaks down the insect's bodies so that they may be absorbed as plant food. The following are few examples of insectivorous plants Dionaea (Venus's Flytrap), Nepenthes, Sarracenia (Pitcher Plant), Darlingtonia (California Pitcher Plant) and Utricularia (Bladderwort). (Go see.)

Inserted Attached, as a stamen on a corolla.

Insertion The place or manner of attachment, as the insertion of petals.

In-situ Being in the original position. Many experiments are conducted in-situ, or in place in the field.

Insoluble This occurs when molecules in a solid material are put into a solvent and are unable to leave the solid and enter the solvent

Instar One of the stages of growth between molts, in the life history of certain insects.

Intake rate The rate at which rain or water enters the soil. Expressed in inches per hour.

Integral membrane proteins Proteins embedded within the lipid bilayer of cell membranes.

Integrated control Integrated use of both biological and chemical methods for controlling pests and weeds

Integrated pest management (abbr. IPM) A combination of strategies to reduce losses due to pests and pathogens based on environmental and economic considerations.

Integrin A transmembrane protein that mediates the adhesion of cells to the extracellular matrix.

Integument The outer envelope of an ovule, which, with other parts, forms the seed coat.

Integuments Two distinct tissue layers that surround the nucellus of the ovule When mature they become the testa or seed coat

Intense selection Selection for the trait(s) of interest with little or no retention of lines not having the trait(s) of interest.

Intensive farming A system of farming with the aim to produce the maximum number of crops in a year with a high yield from the land available and to maintain a high stocking rate of livestock.

Inter Between or among.

Interaxillary Situated between or within the axils of leaves.

Intercalary meristem A meristem found between already differentiated tissues. Occurs in the internodes of grasses (just above leaf nodes) and enables longitudinal growth of the stem.

Intercalating dye A dye molecule which can insert between bases of DNA. Used in laboratories to detect DNA sequences.

Intercalation When a molecule, such as Ethidium Bromide or SYBR Green inserts in between DNA nucleotides of a double-stranded DNA moledule. Ethidium Bromide and SYBR Green are therefore, used in detecting the presence of DNA.

Intercellular Between or among cells.

Intercrop To grow two or more crops simultaneously on the same area of land.

Intercropping The growing of more than one species on the same piece of land at the same time

Interdisciplinary Of, relating to, or involving two or more academic disciplines that are usually considered distinct.

Interfoliaceous Situated between the leaves, especially opposite leaves. Also interfoliar.

Intermediate filament A cytoskeletal filament about 10 nm in diameter that provides mechanical strength to cells in tissues.

Internal transcribed spacers Regions of a DNA sequence that separate genes for certain ribosomal RNAs (rRNAs). These genes appear to be nearly identical among a wide variety of species, but the DNA spacers between them are quite variable and may be species-specific. This makes them good potential targets for "primers" for polymerase chain reaction (PCR) amplification that is used to identify a wide variety of organisms.

Internode The part of stem that lies between two nodes or joints on a plant.

Interphase The stage of cell division when DNA is replicated; a copy of each chromosome is made.

Interveinal Between (leaf) veins .

Intervention Under the EC price support mechanism for agricultural produce was set up in 1972. Its functions included licensing trade in a range of agricultural products outside the EC, intervention buying and the operation of other market support arrangements. Intervention buying or support buying was designed to ensure that farm incomes were not unduly suppressed by over production within the EC. When the free market price of specified commodities falls someway below the target price, the board intervenes in the market, by purchasing and storing the produce. The commodity is then subsequently resold within or outside the EC, disposed of as food aid or destroyed.

Intra Within or inside.

Intracellular signal transduction A chain of reactions that transmits chemical signals from the cell surface to their intracellular targets.

Intracellular Through or within cells.

Intrazonal A term used to describe a soil group with characteristics determined by local factors, such as parent material.

Intron A portion of DNA located between coding regions in a gene which is transcribed, but spliced out of the mRNA before translation. This region of nucleotides does not actually code for any proteins.

Introrse Turned or facing inward, such as an introrse anther facing toward the center of the flower.

Intumescence (edema or oedema) Blisters produced on leaves and other plant parts under conditions of high moisture and restricted transpiration .

Invertebrate An animal without a backbone. Butterflies are invertebrates.

Invertebrates Animals without backbones

Inverted Upside down.

Involucel A secondary involucre in a compound cluster of flowers, as in many of the Umbelliferae.

Involucral bracts A series of bracts beneath or around a flower or cluster of flowers

Involucral Pertaining to an involucre.

Involucrate Having an involucre.

Involucre A collection of whorls of small leaves or bracts beneath a flower or an inflorescence.

Involucriform Resembling an involucre.

Involute Rolled inward, as the edge of a leaf.

Ion An atom or a compound with a net (positive or negative) electrical charge.

Ion channel A protein that mediates the rapid passage of ions across a membrane by forming open pores through the phospholipid bilayer.

Ion pump A protein that couples ATP hydrolysis to the transport of ions across a membrane.

Ions Molecules with a positive or negative charge

Iron chelate A form of iron that is readily available to plants when iron in the soil isn't available to the roots. Products that contain iron chelates can be added to the soil or sprayed on the foliage.

Iron oxides Group name for the oxides and hydroxides of iron. Includes the minerals goethite, hematite, lepidocrocite, ferrihydrite, maghemite, and magnetite. Sometimes referred to as "sesquioxides," or "iron hydrous oxides."

Iron Plants need this soil nutrient to manufacture chlorophyll. In alkaline soil, plants may not be able to absorb iron from the soil, which causes the leaves to turn yellow with green veins. Products, containing iron chelates can be added to the soil or sprayed on the plant's leaves to make iron available to the roots and foliage.

Irregular Not having all the members of the same part alike; asymmetrical; lacking uniformity (often refers to flowers).

Irregular flower A flower with bilateral symmetry.

Irrigate To pass a liquid over or through; specifically, to water by causing a stream or streams to flow over land.

Irrigation The act of watering or moistening; the covering of anything with water or other liquid for the purpose of making or keeping it moist, such as the distribution of water over land to promote the growth of plants.

Isadelphous Having the separate bundles of stamens equal in number, as in the flowers of beans.

Isandrous Having the stamens similar and equal in number to the divisions of the corolla.

Isanthous Having regular flowers.

Island arc Series of volcanic islands that are built above a subduction zone.

Isogamous Having flagellated gametes that are identical in size and structure.

Isogynous Having the pistils, or the carpels of which the single pistil is composed, equal in number to the sepals.

Isolate A culture or subpopulation of a microorganism separated from its parent population and maintained in some sort of controlled circumstance;

To remove from soil or host material and grow in pure culture.

Isomerous Composed each of an equal number of parts, as the members of the several circles of a flower; exhibiting isomerism.

Isometric Usually used for virus particles to describe those that are icosahedral in structure and appear approximately round.

Isomorphic alternation of generations gametophyte and sporophyte generations are indistinguishable except by their sexual organs.

Isoprenoids Also known as terpenes, this is a diverse class of biomolecules. They have in common a five carbon branched molecule as the basic building block.

Isostemonous Having the stamens equal in number to the sepals or petals.

Isostemony The state of being isostemonous.

Isotope One of multiple forms of an element that has a different number of neutrons than other atoms of that element. Some elements have isotopes that are unstable or radioactive, while others have "stable isotopes." Stable isotopes are not radioactive, and do not decay over time. For example, most nitrogen atoms have 14 neutrons, while a very small percentage of naturally-occurring nitrogen atoms have 15 neutrons. These ^{15}N atoms are referred to as stable isotopes. It is possible to use special equipment to determine how much of a stable isotope is present in something. For example, scientists can take a leaf sample and determine how much of the nitrogen present in the leaf has 14 neutrons and how much has 15 neutrons. Because stable isotopes do not break down, and because scientists can measure exact quantities present in samples, it is possible to use stable isotopes as tracers.

Isozyme An enzyme that is similar in structure and function to another enzyme.

ITS (acronym for internal transcribed spacers) Regions of a DNA sequence that separate genes for certain ribosomal RNAs (rRNAs). These genes appear to be nearly identical among a wide variety of species, but the DNA spacers between them are quite variable and may be species-specific. This makes them good potential targets for "primers" for polymerase chain reaction (PCR) amplification that is used to identify a wide variety of organisms.

J

Janus kinase (JAK) A family of nonreceptor protein-tyrosine kinases associated with cytokine receptors.

June census Annual farm returns on the 4th June each year provide the basis of MAFF published statistics regarding holding size enterprises, labour etc.

Juvenile An immature form that appears similar to but usually smaller than the adult and is not sexually mature (e.g. insects with gradual metamorphosis, nematodes).

K

Kainite A natural salt contained in a fertilizer that is used as a source of potash (14% k).

Kaolin minerals A group of nonswelling clay minerals in which one layer of silicon and oxygen alternates with a layer made up of aluminum, oxygen, and hydrogen.

Karyogamy Fusion of the nuclei of two gametes; part of syngamy.

Karyokinesis (karyokineses) Nuclear division in mitosis or meiosis.

Karyotype A photograph of an individual's chromosomes arranged in order showing the number, size, and shape of each individual chromosome.

Kcat The number of reactions per second that the enzyme can catalyze under optimum conditions.

Keel 1. A central ridge along the back of any organ of a plant.

2. The lowest, fused petals of a butterfly-shaped flower.

Keiki A vegetative offshoot (plantlet) formed at a node, as on an inflorescence (of an Orchid).

Keratin A type of intermediate filament protein of epithelial cells.

Kernel 1. The inner portion of a seed, usually refers to an edible seed or the central part of a nut or fruit seed.

2. The whole seed of a cereal such as corn or wheat.

3. To form kernels.

Key management species Plant species on which management of a specific unit is based.

Kilobase (kb) One thousand nucleotides or nucleotide base pairs.

Kinesin A motor protein that moves along microtubules towards the plus end.

Kinetochore A specialized structure consisting of proteins attached to a centromere that mediates the attachment and movement of chromosomes along the mitotic spindle.

Kinetochore microtubules Microtubules of the mito-tic spindle that attach to condensed chromosomes at their centromeres.

Km The Michaelis constant. It is an indication of the affinity of the enzyme for the substrate.

Knee 1. A sharp bend in a tree branch

2. A protruding enlargement of the roots at the base of certain swamp-growing trees.

Kneeler A device to help gardeners lower themselves down or push themselves up from a kneeling position. It usually has handholds several feet high attached to a cushioned platform. It may be reversed and used as a garden bench.

Kneeling pads A usually rectangular and waterproof mat that is used for cushioning the knees while gardening.

Kneepads Cushions that are strapped behind the knees to protect the kneecaps while kneeling or crawling.

Knockout genes Transgenes designed to silence an existing gene.

Knock-outs A plant in which a gene has been rendered non-functional. By eliminating the gene's function, the investigator can infer its role in the plant's physiology from the phenotype.

Knot A lump or swelling in or on a part of a plant, such as the node of grass, or the thick part of the wood of a tree where a branch grows out of the trunk.

Knurl A thick, deformed knot in wood.

Koch's postulates The procedure used to prove the pathogenicity of an organism, i.e. its role as the causal agent of a disease.

Krebs cycle See citric acid cycle.

Krill shrimp-like planktonic crustaceans; an abundant and major food source for Antarctic animals such as whales, fish, and penguins.

L

L.s. Longitudinal (lengthwise) section.

Labellum A lip, particularly the lip of an orchid. The median, modified petal of an Orchid flower.

Labiate Two-lipped, as a calyx or corolla divided into two unequal parts, like the lips of a mouth.

Labile Readily undergoing change or breakdown. The term labile is often used to describe compounds (such as carbon or nitrogen) that are available to easily move through soils.

Labile phosphorus Intermediate form that is rather weakly adsorbed or bound to various compounds and clay in the soil.

Labium The lower lip of a flower with two lips.

Laccate Appearing as if varnished.

Lacerate Appearing torn as the edge of a sepal or petal; irregularly cut or cleft.

Lacinia A long slash or incision in a leaf, petal, or similar organ; one of the narrow lobes resulting from such cuts.

Laciniate Irregularly cut into narrow-pointed lobes; fringed; refers to petals, bracts, leaves, etc. Also laciniated, laciniose.

Lacinula A small, narrow lobe or fringe.

Lacinulate Having small, finely cut leaves or petals.

Lacrimiform Teardrop-shaped.

Lacuna A hole or cavity.

Lacunulose Having tiny holes.

Lacustrine deposit Material deposited from lake waters.

Lagging strand The strand of DNA synthesized opposite to the direction of movement of the replication fork by ligation of Okazaki fragments.

Lamella A thin plate, or an organ or part resembling one. Raised, membranous outgrowths on the labellum, as in the group, Coelogyne. Also lamel.

Lamellate Composed of or having lamellae. Also lamellar.

Lamellipodium A broad, actin-based extension of the plasma membrane involved in the movement of fibroblasts.

Lamina A thin plate or scale. Specifically, the blade or expanded part of a leaf or petal.

Laminin The principal adhesion protein of basal laminae.

Lamins Intermediate filament proteins that form the nuclear lamina.

Lanate Woolly; covered with long, fine, curly hairs.

Lance-linear Long, thin, and pointed. (i.e. a lance-linear leaf)

Lanceolate Long and thin and broadest below the middle, tapering to a point like a lance; lance-shaped.

Lanciform Lance-shaped; long, narrow and pointed, sometimes broader at the base than at the tip.

Land 1. Refers to the solid surface of the earth.

2. A broad term that refers to soil, surface soil, mineral deposits, and water supplies in a certain area.

Land capability classification Soil that is grouped into special units, subclasses, and/ or classes according to their capability for intense use and treatments that are required for sustained use.

Land race Plant stocks selected by farmers on a local basis over many years.

Landform A geomorphological term used to express earthly surface formations.

Landscape A view or prospect of rural scenery, more or less extensive, such as seen from a single point of view.

Landscape architect A person who designs and often supervises the placement of elements (plants, footpaths, etc.) in an outdoor setting.

Landscape fabric Any of various porous organic or inorganic cloths or mats used to discourage weeds from growing. Also called weed mat.

Landscape gardening The art of laying out grounds and arranging trees, shrubs, flowers, paths, fountains, etc. to produce an attractive surrounding.

Land-use history Referring to how an area was used in the past.

Lanose Woolly; lanate.

Lanuginous Woolly or downy; covered with soft fine hairs.

Lanugo Soft, thick downy growth.

Larva Immature stage of certain animals (especially insects) that undergo complete metamorphosis; the wormlike or caterpillar stage of the life cycle of such insects.

Latent Present but not manifested or visible, as a symptomless infection by a pathogen.

Latent Infection Where the host is infected with a pathogen but does not show any symptoms.

Latent period The time between infection and the production of new inoculum; the time after a vector has acquired a pathogen and before it can be transmitted.

Lateral A side shoot that grows from a shoot or root; side growths, which develop from the main branches of a tree, shrub, or other plant.

Lateral bud (axillary bud) Bud that develops in the axil of a leaf.

Lateral Flow Strip A variation on the ELISA tests, using strips as opposed to microtiter wells to detect the presence of a protein produced.

Lateral lobes The two lobes to either side of the midlobe of a three-lobed labellum.

lateral moraines An accumulation of soil, pebbles, rocks, and boulders distributed and marking the lateral edges of a glacier or ice sheet.

Latex A milky, usually white or yellowish fluid produced by plants, such as Asclepias and Euphorbia.

Lath house A framework of widely spaced narrow boards used to protect plants from the sun by providing partial shade.

Laticiferous Bearing, containing, or secreting latex.

Latitude The angular distance north or south from the Earthís equator, measured in degrees, minutes, and seconds. One degree equals 60 nautical miles. A combination of latitude and longitude can be used to locate any spot on the earthís surface.

Lava Melted, or molten, rock is called magma if it is below the earth's surface, and lava if it reaches the earth's surface.

Lawn Space, usually level, covered with grass and kept smoothly mown, such as in front of a home, etc.

Lax Loosely or widely spaced, as a lax cluster of flowers.

Layering A method of propagation in which a shoot of a living stem is bent and pegged down to the soil. The shoot develops roots while still attached to the parent plant; once established, it is cut from the mother plant to form a separate plant, which can then be potted up. The illustration shows a shoot bent down and forming roots below the soil's surface. A stake supports the end of the shoot.

LD50 The dose of chemical that can kill 50% of the test subjects.

Leach To wash or drain by trickling water through; to treat by downward drainage, as rain washing away nutrients from the soil.

Leaching The process in which nutrients and other compounds (e.g., calcium) in the upper layers of soil are dissolved and carried to lower layers and, in some cases, to groundwater.

Leader The main or terminal shoot of a plant, such as the topmost point of a fir or larch tree or the strongest shoot of a young fruit tree.

Leading strand The strand of DNA synthesized continuously in the direction of movement of the replication fork.

Leaf An often flattened photosynthetic organ attached to, and produced by, a stem.

Leaf area index (LAI) The area of green leaf per unit area of ground

Leaf dip Inspection of a sample of sap from a suspected virus-infected leaf using an electron microscope.

Leaf gap An interruption of a siphonostele located above the point of attachment of a leaf trace.

LEAF Linking Environment and Farming

Leaf These grow along the sides of branches or stems or at the bases of the plants. Most are green and contain chlorophyll, though they vary in their shapes and sizes. They are the part of the plant that ordinarily performs photosynthesis, the process that converts sunlight and carbon dioxide into energy. It is common for leaves to grow, fall off, and be replaced. Plants entering dormancy may lose all their leaves. Maples, for example, drop their leaves in autumn and replace them in the spring and summer.

Leaf mold This is a compost of decayed leaves. It is made by piling leaves of deciduous trees and shrubs together and allowing them to rot. The pile should be trodden firm and the leaves should be wet. Every 6 months or so, the pile should be turned over so the outer section of the pile is in the center. It takes a year and a half to two years for the leaves to decompose enough to use for leaf mold.

Leaf plant A plant grown for its decorative foliage.

Leaf scar A mark left on the stem where a leaf has been lost or removed. It can easily be seen on Palms where old leaves have been shed.

Leaf scorch This may be caused by dryness, high temperatures, disease, or lack of certain elements in the soil. The foliage looks as if its been dried by heat.

Leaf sheath Lower, tubular part of a grass leaf that clasps the culm.

Leaf spot A plant disease lesion typically restricted in development in the leaf after reaching a characteristic size.

Leaf trace One or more vascular bundles extending from the stele of a stem into a leaf.

Leafbud A bud producing a stem and leaf, unlike a flowerbud, which contains a blossom.

Leafhopper Mobile insects with sucking mouthparts (order Homoptera).

Leaflet One of the separate blades or divisions of a compound leaf.

Leafroll Disease symptom in which the edges of the leaf roll or turn up; often a symptom of virus infection.

Leafstalk The stalk that attaches a leaf to the stem or branch; the petiole.

Legume Simple, dry, dehiscent fruit developed from a simple pistil and splitting at maturity along two seams;

Legumes plants of the family leguminosae, e.g. clover, peas

Lemma The lower of two bracts that protect the floret in grasses.

Lenticel A lens-shaped body of cells, formed on the outside of a woody plant stem, which serves in the exchange of gases between the stem and the outer air.

Lenticellate Pertaining to or having lenticels.

Lenticular Resembling a lentil or a double-convex lens, as some seeds.

Lentiginous Covered with minute dots; freckled; speckled.

Lepidote Covered with dry scales or scaly spots.

Leprose Having scaly or dry scales.

Leptoid A food-conducting cell of a moss.

Leptosporangium (leptosporangia) A meiosporangium that arises from a single initial cell and forms a wall of a single layer of cells.

Leptotene The initial stage of the extended prophase of meiosis I during which homologous chromosomes pair before condensation.

Lesion Localized diseased area or wound.

Leucine zipper A protein dimerization domain containing repeated leucine residues; found in many transcription factors.

Level To prepare soil for planting by raking gently.

Level terrace An embankment, constructed on a contour, which cuts horizontally across the slope. A means of creating a level area on hillsides or slopes.

LFA Less Favored Area

Liana A climbing and twining, usually woody, plant found mostly in tropical rainforests, which may climb to the level of the tree canopy. Also liane.

Lid The cap of a box-like seed capsule.

Life cycle A representation of all the stages of an organisms life from birth through reproduction.

Ligand A molecule that binds to a receptor.

Ligase An enzyme that can rejoin cut strand of DNA.

Light reactions The reactions of photosynthesis in which solar energy drives the synthesis of ATP and NADPH.

Light soil A term used to indicate a sandy or coarse-textured soil that drains rapidly. It also may lack important plant nutrients.

Ligneous Consisting of or resembling wood; woody, as distinguished from herbaceous.

Lignescent Tending to be or become ligneous or woody; somewhat woody.

Lignification Hardening of tissue through the deposition of lignin in the cell wall.

Lignify To become wood or woody.

Lignin A complex polymer deposited in some cell walls of vascular plants. Lignin is one of the main constituents of secondary walls and wood. It gives compressive strength and rigidity to the cell wall as well as making it impermeable to water.

Ligniperdous Destructive of wood.

Lignituber A structure formed by the deposition of lignin around the tip of a fungal hypha penetrating a host plant cell. The formation of a lignituber is an active defense mechanism and its function is to inhibit the growth of the hypha into other cells.

Lignum Woody tissue, as contrasted with soft tissue or with bark.

Ligulate 1. Strap-shaped, such as a corolla, leaf, or petal.

2. Furnished with ligule, as a ligulate grass.

Ligule A strap-shaped plant part; the blade formed by the corolla in some or all the florets of many members of the daisy family, Compositae. The strap-shaped part that sticks up from the top of the sheath in many grasses, palms, and some other plants.

Liguliflorous Having heads composed exclusively of strap-shaped florets.

Limb 1. A primary branch of a tree.

2. The expanded, flat portion of a plant part, such as the upper portion of a petal.

Limbate Bordered (i.e. a flower in which one color is edged with another color).

Lime Compounds of calcium (ground limestone-calcium carbonate, hydrate limestone-calcium hydroxide, or burnt lime-calcium oxide/CaO). The amount of lime in the soil determines whether it is acid, neutral or alkaline. Limestone that is worked into the soil raises the pH from an acid pH to one that is alkaline. When lime is dug into stiff or clayey soil, it tends to make it friable and more easily worked. On the other hand, loose, sandy soil that dries up quickly is made more compact and retentive of moisture by the addition of lime. Many diseases, such as club root and Clover sickness, come from the soil and attack certain crops; lime prevents some of these diseases. Furthermore, lime is distasteful, if not actually harmful, to some soil pests, such as slugs, wireworms and leatherjackets.

Lime requirement The amount of standard ground limestone required to bring a layer of an acre of acid soil (6.6 pH) to some specific lesser degree of acidity. Expressed in tons per acre of nearly pure limestone.

Lime sulfur A caustic pesticide made from calcium polysulfide. Usually used as a dormant spray to control various diseases, such as leaf spot, peach leaf curl, and powdery mildew. Also effective in controlling mites, scale, and some other insects.

Limnology The scientific study of the life and phenomena of fresh water, especially lakes, ponds, and streams.

Line Plants within a species that have the same genetic composition and are genetically pure, (i.e., inbred line). Lines are typically not agronomically competitive and are used only in plant breeding.

Linear Long and very narrow with parallel sides, such as a linear leaf. Can also be used in combination, as a linear-lanceolate leaf, which is a leaf that is long, narrow and pointed.

Lineate Marked with lines. Also lineated, lined.

Lineolate Marked with fine lines. Also lineolated.

Lingy Abounding in heaths; heathy; heathery.

Linkage The association of genes on the same chromosome. The shorter the distance between two genes, the greater the probability they will be inherited together.

Linkage disequilibrium The condition whereby alleles at two loci occur together in a population at a greater frequency than that predicted by the product of the individual allele frequencies.

Linkage map An ordered representation of genes on a chromosome derived from observed percentages of recombination. Units on a linkage map are centiMorgans (cM), with 1 cM approximately equal to 1% recombination.

Linkage drag The inheritance of undesirable genes along with a beneficial gene due to their close linkage.

Lip 1. One of the parts of a bilabiate (two-lipped) flower.

2. In orchids, it is one of the three petals that has a different shape than the other two.

Lipid bilayer Made up of compounds which have both hydrophobic and hydrophilic moieties, this is the structure of membranes such as the cellular membrane.

Lipid synthesis inhibitors Mode of action of herbicides that target the production of lipids which are essential in plant growth and function.

Lipids Hydrophobic molecules that function as energy storage molecules, signaling molecules, and the major components of cell membranes.

Lipophilic Having chemical properties relating to lipids (lipid-like), nonpolar compounds that are highly soluble in organic solvents, but not water; hydrophobic.

Lithophyte A plant that grows on the surface of rocks, getting its nourishment from the air, as some orchids.

Lithosol A shallow soil consisting of imperfectly weathered rock fragments or nearly all barren rock.

Lithosphere A strong layer of the Earth, about 100 km in thickness, consisting of the Earth's crust and a portion of the underlying upper mantle. This layer lies on top of the weaker asthenosphere.

Litterfall Leaves, twigs, and other plant material that falls to the ground.

Littoral Pertaining to a shore, especially of the sea; growing on or near the shore.

Lituate Forked, with the points turned outward.

Liturate Spotted.

Local lesion Small, restricted lesion, often the characteristic reaction of differential cultivars to specific pathogens, especially in response to mechanical inoculation with a virus.

Locule a chamber in a sporangium or ovary.

Locus The site on a chromosome where a particular gene (or other sequence) is located.

LOD Abbreviation for logarithm of odds, a significance measure used in maximum likelihood statistics. In interval mapping, the LOD score is log10 of the ratio of the probability that a QTL is present to the probability that a QTL is absent. For example, if the presence of a QTL is 1000 times more probable than its absence, then the LOD score = 3, because log10 of 1000 = 3.

Lodge To fall over.

Lodged Occurs when plant stalks (such as corn plants) have been damaged by insect, disease, or climate which results in the plant falling over.

Lodging Term used when the crop canopy becomes flattened by heavy rains. Often triggered by severe weather at the soft dough stage. The problem is a combination of heavy heads and weak stems. Modern breeding increases stem strength.

Lodicules Two small structures below the ovary which, at flowering, swell up and force open the enclosing bracts, exposing the stamens and carpel

Loess Material transported and deposited by wind and consisting of predominantly silt-sized particles.

Log Kow Octanol/water partition coefficient, relative measurement of how water soluble a chemical is. A compound with a low log Kow would be more water-soluble than a compound with a high log Kow.

Long terminal repeat (LTR) DNA sequences found at the ends of retroviral and retrotransposon DNA that are direct repeats of several hundred nucleotides resulting from reverse transcriptase activity.

Longitude The angular distance east or west from a meridian drawn between the North Pole to the South Pole and passing through Greenwich, England, measured in degrees, minutes, and seconds. A combination of latitude and longitude can be used to locate any spot on the Earthís surface.

Lorica A vase-shaped or cup-shaped outer covering. Found in many protists, including some flagellates, ciliates, chrysophytes, and choanoflagellates, as well as in some animal cells.

Lowland Farming Farming on low lying land as distinct from uplands, usually more intensive

LS Longitudinal Section. The tissue sections were cut along the long axis of the grain. The central sections go lengthwise through the embryo and along the length of the ventral groove

Lumen Central cavity of a cell or other structur.

Lymphocyte A blood cell that functions in the immune response. B lymphocytes produce antibodies and T lymphocytes are responsible for cell mediated immunity.

Lymphoma A cancer of lymphoid cells.

Lyrate Resembling a lyre; for example, a lyrate leaf is one that is divided into several lobes, which increase in size toward one large terminal lobe.

Lysimeter Devices that are placed in the ground to collect soil water. There are two types of lysimeters in use: tension, and zero-tension. Zero-tension lysimeters collect soil water that is naturally (i.e., only gravity is influencing) percolating downward through soils. Tension lysimeters have a vacuum (similar to what roots might exert on soils) applied to them and gently "suck" soil water through a porous material. Zero-tension lysimeters are designed to capture soil water that might otherwise make its way into groundwater or lower soil horizons, while tension lysimeters are designed to capture soil water that roots might take up.

Lysine Nonessential amino acid found in cheese, eggs, milk, meat, legumes, whole grains, and nuts

Lysis the dissolution or destruction of cells by the disruption of the cell membrane.

Lysogeny A viral infection leading to integration of an inactive copy of viral DNA into the cell genome.

Lysosome Eukaryotic organelle which carries digestive enzymes. The lyzosome fuses with a vacuolar membrane containing ingested particles, which are then acted upon by the enzymes.

M

M phase The mitotic phase of the cell cycle.

Macerate To cause disintegration of tissues by separation of cells; to soften by soaking.

Macroconidium The larger of two kinds of conidia formed by certain fungi .

Macrocyclic A rust fungus that typically exhibits all five stages of the rust life cycle.

Macromolecules Large molecules which are composed of smaller individual molecules.

Macronutrient An element needed in relatively large quantities for plant growth, e.g. nitrogen (N), phosphorus (P), and potassium (K); .

Macronutrients Elements needed in relatively large amounts for plant growth. Includes carbon, hydrogen, sulfur, nitrogen, phosphorus, potassium, calcium, magnesium, and oxygen.

Macrophage A type of white blood cell specialized for phagocytosis.

Macrophyllous Having large leaves, usually with many veins.

Macrophyte Plants large enough to be seen with the unaided eye.

Macropodous Having a long stalk or stem; (of a plant embryo) having a large or long hypocotyl.

Macropore A large, non-capillary pore found in soil through which water drains by gravity.

Macular Spotted.

Maculation Spotting or marking.

MAFF Ministry of Agriculture Fisheries and Food.

Magnesium An element found in combination in minerals; in plants, located either in the seeds or in the chlorophyll.

Maintenance approach Sets a soil test level goal and recommends fertilizer to build the soil to the specific nutrient goal; emphasis is placed on maintaining the soil fertility level at or above the point of the economic maximum yield; one of three approaches to making fertilizer recommendations.

Malathion A man-made insecticide that is effective in destroying a wide range of garden pests, such as aphids, scale and mites.

Male Having flowers that do not produce fruit or seed and only contain stamens.

Malignant tumor A tumor that invades normal tissue and spreads throughout the body.

Malonylation Conjugation of a glucose conjugate of xenobiotic to malonyl in the presence of malonyl-CoA and malonyl-CoA-transferase; considered a Phase III reaction or secondary conjugation.

Mammillate Having nipple-shaped protuberances.

Mantle (hyphal sheath) Dense hyphal mass of ectomycorrhizal fungus enclosing the short feeder roots of plants.

Mantle The 2900 km thick zone of the Earth below the crust and above the mantle made up of minerals that contain iron, silica, magnesium and oxygen.

MAP kinases A family of mitogen-activated protein-serine/threonine kinases that are ubiquitous regulators of cell growth and differentiation.

Marcescent Withering, but not falling off. (Refers to a plant part.)

Marginal farmland 1. Surface soil that is only a few inches deep.

2. Land repeatedly farmed without benefit of humus or chemical replacements.

Marker An easy to detect trait controlled by a known gene. Markers, such as antibiotic or herbicide resistance, are often used to determine if an organism is transgenic.

Marker gene A gene controlling an easily detectable trait used to select transgenic cells or plants from non-transgenic ones, (i.e., herbicide and antibiotic resistance).

Marker-assisted selection A procedure used in plant breeding in which genetic markers that are (a) easy to identify and (b) linked to desirable genetic traits that are difficult to identify (such as disease resistance) are used to aid in selection from a population.

Marl An earthy deposit consisting mainly of calcium carbonate; used for liming acid soil.

Marrow 1. The pith of plants.

2. The pulp of fruits.

Mass flow Process by which a dissolved or suspended molecule is moved with movement of the solvent, i.e. a floating piece of wood moving with the current in a stream.

Mass wasting Movement of disintegrated rock materials by sliding or falling to the base of a slope by the force of gravity.

Master horizon A principal horizon, such as the A, E, B, C, or R.

Mastigoneme Small hair-like filaments found on the "hairy" flagellum of the Chromista.

Maternal inheritance (also cytoplasmic or extrachromosomal inheritance) Inheritance of non-nuclear genes, i.e. those in mitochondria and chloroplasts.

Mating types Compatible strains, usually designated + and - or A and B, necessary for sexual reproduction in heterothallic fungi.

Matrix The inner mitochondrial space.

Maturation promoting factor (MPF) A complex of Cdc2 and cyclin B that promotes entry into the M phase of either mitosis or meiosis.

Maximum Contaminant Limit (MCL) The highest level of a contaminant that is allowed in drinking water.

Meadow 1. A piece of low-lying, moist land that is often used for growing hay or as pastureland.

2. An upland area covered with grass and usually surrounded by woods.

Meal Consists of canola seed remains after oil has been extracted from them.

Mean The average value of a set of numbers

Mechanical analysis The physical analysis of soil material to determine its grain-size fraction.

Mechanical injury Injury of a plant part by abrasion, mutilation, or wounding.

Mechanical stability The resistance of soil broken down by mechanical forces like tillage or abrasion.

Mechanical transmission Spread r introduction of inoculum to an infection court (wounding) by human manipulation, accompanied by physical disruption of host tissues.

Mechanism of action The specific process inhibited by a herbicide.

Medicinal plants Plants that are or have been used medicinally.

Medium A mixture of organic and/or inorganic chemical compounds and water that provides the nutrients needed for the growth of a microorganism in vitro; for higher plants, a mixture of fertilizers and other components in which a plant is growing.

Medium milk A description the half-grown grain. When punctured the grain contents are milky.

Medium sand Soil particles measuring between 0.25 to 0.50 millimeters.

Megabase (Mb) One million nucleotides or nucleotide base pairs.

Megasporangium (megasporangia) -A sporangium that produces megaspores.

Megaspore A meiospore that forms a female gametophyte.

Megasporocyte A cell that undergoes meiosis to form megaspores.

Megasporophyll A sporophyll that bears megasporangia.

Meiocyte Any diploid cell that undergoes meiosis.

Meiosis (meioses) A form of cell division of eukaryotes in which each of the four daughter cells has half as many chromosomes as the mother cell; produces meiospores in plants and gametes in animals.

Meiospore A spore produced as a result of meiosis.

Meiosporocyte The diploid cell that undergoes meiosis to produce meiospores.

Melanin Brown-black pigment; common in sclerotia and other survival structures.

Membranaceous Thin, soft and translucent.

Membrane Semi-fluid structure which bounds all cells, and partitions the interior of eukaryotic cells. It consists primarily of two lipid layers, with proteins "dissolved" in the lipids.

Membrane-anchored growth factors Growth factors associated with the plasma membrane that function as signaling molecules during cell-cell contact.

Mentum A "chin" on some Orchid flowers, formed by the fusion of the lateral sepals and the column-foot.

Merchantable Suitable for sale; "merchantable timber."

Mericarp One of the halves of the fruit of certain plants, such as the maple, whose fruits split into halves.

Meristem Plant tissue characterized by frequent cell division, producing cells that become differentiated into specialized tissues.

Meristem culture Aseptic in vitro culture of a plant or plant part from a portion of the meristem; a method used to produce pathogen-free plants.

Mesic- Neither especially dry nor especially wet; applied to features of plants or places where they grow.

Mesocarp The central portion of the ripened wall of a fruit.

Mesochile The middle part of the lip in certain orchids that have this organ divided into three parts.

Mesocotyl The region where vascular tissues of the root join the epicotyl at the scutellar node

Mesoderm The middle germ layer; gives rise to connective tissues and the hematopoietic system.

Mesokaryotic Nuclear condition unique to the dinoflagellates in which the chromosomes remain permanently condensed.

Mesophyll The photosynthetic tissue of a leaf, located between the upper and lower epidermis. Mesophyll is commonly differentiated into palisade parenchyma and spongy parenchyma.

Messenger RNA (abbr. mRNA) A form of RNA that carries information to direct the synthesis of protein.

Metabolic Pathway A series of chemical reactions catalyzed by enzymes in a living system. The products of the pathways have vital functions. For example, the Kennedy metabolic pathways found in plants are involved in producing various types of seed oil.

Metabolism A general term referring to the change of a herbicide from an active to an inactive state.

Metabolite Any chemical participating in metabolism; a nutrient.

Metal An element, such as iron or copper, that are good conductors of heat and electricity, and have a characteristic luster.

Metal sulfide A mineral in which the element sulfur is linked with a metal, such as iron sulfide in the mineral pyrite, commonly known as fool's gold.

Metamorphic rock A rock derived from a pre-existing rock that has been changed in its minerals, chemistry and structure in response to changes in temperature, pressure and stress.

Metaphase The phase of mitosis during which the chromosomes are aligned on a metaphase plate in the center of the cell.

Methionine Essential amino acid found in cereal, whole grains, sesame and sunflower seeds, and yeast

Micro nutrients Very small amounts of nutrients needed by plants.

Microarray A small glass, or filter, square containing probes for thousands of gene products.

Microbe A microscopically visible organism in the size range of 1/100 to 1/1000 millimeter, mostly consisting of only a single cell. Individual Bacteria and Archaea are microbes.

Microbes Microorganisms. Soil microbes decompose dead plant and animal biomass.

Microbial ecology The branch of ecology that deals with the study of the interactions of living microorganisms with one another and with their nonliving environment of matter and energy.

Microbial Pertaining to or relating to microbes, or microorganisms.

Microclimate A small, local climate within a larger climate area, such as a greenhouse or a protected area of a garden.

Microconidium The smaller of two kinds of conidia formed by certain fungi .

Microcyclic Describing a rust fungus that produces only teliospores and basidiospores.

Microfibers A method used to transform tissue culture cells using tiny fibers coated with DNA. These fibers are combined in solution with the cells and shaken vigorously causing the fibers to stab the plant cells delivering the DNA to them.

Microfibrils In guard cells, microfibrils are made of cellulose and are oriented transversely to the long axis of the cell resulting in cell expansion in the direction of its long axis because the cellulose reinforcement offers the least resistance at right angles to its orientation

Microfilament A cytoskeleton filament composed of actin.

Microflora The combination of all microorganisms in a particular environment.

Micron (micrometer) One millionth of a meter.

Micronutrients Plant nutrients needed in very small amounts; generally considered to include: zinc, iron, chlorine, copper, manganese, boron, and molybdenum.

Microorganism (microbe) An organism of such small size that it can only be seen as an individual organism with the aid of microscope.

Microphyll The flattened, photosynthetic leaf-like organ of the Lycophyta.

Microphyllous 1. Having small leaves.

2. Having leaves with only one unbranched vein.

Micropropagation Tissue culture technique that asexually propagates a clone of a plant from tissue

Micropyle the opening in the integument of the ovule of a seed plant through which the pollen grains or pollen tubes pass.

Microsclerotium Microscopic, dense aggregate of darkly pigmented, thick-walled hyphal cells.

Microsporangium In plants that produce two types of spores, the receptacle that contains only the smaller type; comparable to the pollen sac of the anther in a seed plant.

Microspore The smaller type of spore produced by plants that bear two types.

Microspores Immature pollen grains that can be effectively triggered and cultured under sterile conditions to produce embryos which can be regenerated into plants (a process called androgenesis).

Microsporocyte A cell that undergoes meiosis to form microspores.

Microsporophyll A sporophyll that bears microsporangia.

Microstylous A flower that has short styles.

Microtubule A cytoskeletal component formed by the polymerization of tubulin into rigid, hollow rods about 25 nm in diameter.

Microtubule-organizing center An anchoring point near the center of the cell from which most microtubules extend outward.

Microtubules Type of filament in eukaryotic cells composed of units of the protein tubulin. Among other functions, it is the primary structural component of the eukaryotic flagellum.

Microvilli Thin fingerlike protrusions from the surface of a cell, often used to increase absorptive capacity or to trap food particles. The "collar" of choanoflagellates is actually composed of closely spaced microvilli.

Microvillus An actin-based protrusion of the plasma membrane, abundant on the surfaces of cells involved in absorption.

Middle lamella The layer, consisting largely of pectic substances, between the walls of adjacent plant cells.

Midlobe The middle lobe of a three-lobed labellum, usually brightly colored and furnished with calli (callus), keels, and/or lamelle.

Mid-ocean ridge system The volcanic mountain chain that is formed by volcanic activity at the boundary between two tectonic plates that are drifting apart. Molten rock from the mantle below rises to the surface and solidifies to form new oceanic crust.

Midrib Central, thickened vein of a leaf.

Midvein Runs down the center of the leaf blade; is part of the framework of the leaf and gives support

Migratory Migrating from place to place on a plant or from plant to plant when feeding .

Mildew 1. A growth (usually whitish colored) produced on living plants by fungus.

2. A fungus that causes such growth.

Mineral 1. An inorganic substance.

2. The solid homogeneous crystalline chemical element that results from the inorganic process of nature.

Mineral deposit Accumulation of naturally occurring minerals.

Mineral soil Soil that consists of less than 30% organic matter and is mostly derived from rock.

Mineralisation The conversion of bound organic nitrogen into the mineral (ionic or inorganic) form required for plant uptake. The inorganic nitrogen is then converted in a process known as nitrification from ammonium salts to nitrites, and from nitrites to nitrates by nitrifying bacteria which obtain their energy from this oxidation process. Once free nitrate is formed, the rapid recycling process offers many options, the nitrate may be assimilated by plants, denitrified or leached.

Mineralized The conversion of an element from an organic form to an inorganic state as a result of microbial activity.

Mineralizes The conversion of an element from an organic form to an inorganic state as a result of microbial activity.

Miscible Two solvents that are able to dissolve into each other when mixed

Mismatch repair A repair system that removes mismatched bases from newly synthesized DNA strands.

Mitigation Correction of a problem, often due to the presence of a pollutant.

Mitochondria Cytoplasmic organelles responsible for synthesis of most of the ATP in eukaryotic cells by oxidative phosphorylation.

Mitochondrion (mitochondria) The intracellular organelle that carries out oxidative respiration.

Mitosis This is cell division, such as what occurs after fertilization of the female germ cell by the male. It is usually preceded by halving and differentiation of the chromosomes within the nucleus. Cell division is the process by which plants grow.

Mitotic spindle An array of microtubules extending from the spindle poles that is responsible for separating daughter chromosomes during mitosis. See also kinetochore microtubules, polar microtubules, and astral microtubules.

Mitriform Having a hollow conical shape, resembling a bishop's miter.

Mixed grazing More than one type of animal grazing the same area at the same time

Mixed-function oxidase Enzymes that catalyze oxidation of several substrates.

MLO (mycoplasmalike organsim) Archaic term for phytoplasma; plant-parasitic pleomorphic mollicute (prokaryote with no cell wall) found in phloem tissue; cannot yet be grown on artificial nutrient media.

Model A quantitative representation of the relationships among the entities in a system, often used to make predictions about the system.

Modes of action The method by which an organism is affected, damaged or killed. The mode of action usually involves a chemical reaction instigated by another source. (herbicide, etc.).

Moisture tension The force at which water is held by soil. Expressed as the equivalent of a unit column of water in centimeters.

Mold Any microfungus with conspicuous, profuse, or woolly superficial growth (mycelium and/or spore masses) on various substrates; especially an economically important saprobe; molds commonly grow on damp or decaying matter and on the surface of plant tissues.

Molecular Biology The study of the molecules in living organisms, the reactions they undergo, how these reactions are controlled and how molecules are assembled into larger structures such as membranes and fibers.*

Molecular Farming The application of biotechnology for the selected production of pharmaceutical compounds or other health or industrial compounds within a living organism (e.g. microbe, agricultural crop, livestock).

Molecular marker An identifiable DNA sequence on a chromosome. A marker can be a gene, part of a gene, or a sequence in a non-gene region. SSR, RFLP, RAPD, and AFLP are acronyms for commonly used marker techniques.

Molecular motor A protein that generates force and movement by converting chemical energy to mechanical energy.

Molecule A group of atoms that are bonded together.

Mollicute One of a group of prokaryotic organisms bounded by flexuous membranes and lacking cell walls .

Molt To shed a cuticle or body encasement during a phase of growth.

Molybdenum A trace element present in most soils and plants.

Monadelphous Having the stamens of a flower, united into one group, as can be seen in Hibiscus.

Monander A plant that has only one stamen, such as certain orchids.

Monandrous Having a single stamen.

Moniliform Constricted at intervals and resembling a string of beads, as certain pods, roots and stems.

Monocarp A plant that bears fruit once and then dies.

Monocarpellary A flower consisting of only one carpel.

Monocarpic Bearing fruit once and then dying, as annual plants.

Monocarpous A flower that has a single ovary, whether simple or compound.

Monocistronic Messenger RNAs that encode a single polypeptide chain.

Monoclinous A flower that has both the stamens and pistils.

Monoclonal antibody Antibody produced from clones of a single antibody-producing cell .

Monocot A member of the Division Anthophyta, Class Liliopsida. This class is characterized by having one cotyledon.

Monocots A subclass of Angiosperm plants based on anatomical characteristics.

They tend to have:

-Narrow leaves

-One cotyledon

-Parallel veins in the leaves

-Flower parts are usually in multiples of threes

-A scattered arrangement of primary vascular bundles in the stem

-Fibrous root system

ex. grass, daylilies, corn

Monocotyledon A plant that has a single seed leaf.

Monoculture system of cultivation where only one crop is grown on the same piece of land over a number of years.

Monocyclic Having one disease or life cycle per growing season .

Monoecious Having male and female reproductive organs on a single individual.

Monogastric Non ruminant animal having only one stomach, e.g. pig, human

Monogenic Determined by a single gene .

Monogenic resistance (single gene resistance) Resistance conferred by a single gene .

Monogynoecial Formed from a single pistil of a flower, as in certain fruits.

Monomerous 1. Having only one member in each whorl (pistil, stamen, sepal, or petal), said of a flower.

2. Monocarpellary.

Monopetalous 1. Having only one petal.

2. Having the petals united into one piece by their edges or at the base.

Monophyletic group A group of organisms consisting of a common ancestor and all its descendants.

Monopodia An axis of growth that extends in the direction of previous growth, with lateral branches being produced beneath it, for example, as in certain coniferous trees.

Monopodial Having the growth of the stem or rhizome continuing in the same direction indefinitely. A habit of growth in which new leaves form from the same growing point.

Monopolar Develop or move toward one direction.

Monosaccharides Simple sugars with the basic formula of $(CH_2O)n$.

Monosepalous 1. Having the sepals united.

2. Having only one sepal.

Monosperm A plant that produces only one seed.

Monospermous Having only one seed.

Monosporous Having only one spore.

Monostichous Arranged in a single row along a stem, as certain flowers.

Monostylous Having only one style.

Monosymmetrical Capable of being bisected into similar halves with reference to a single plane; bilaterally symmetrical, as in flower of the sweet pea.

Monotocous Producing fruit only once, as annuals.

Monotypic A group consisting of a single species.

Monovalent cations Ions having a positive charge.

Monoxenic culture A culture containing one species of organism growing in the presence of one other species of organism.

Monstrose This mutation appears as randomly occurring growth from multiple growing points and usually results in a misshapen, club-like plant.

Moor An expanse of open, uncultivated, and more or less elevated land, often overrun with heath.

Morphogenesis The process of morphology development.

Morphology The study of the form of organisms; form and structure of organisms.

Mosaic Disease symptom characterized by non-uniform coloration, with intermingled normal, light green and yellowish patches, usually caused by a virus; often used interchangeably with mottle.

Most limiting nutrient Justus von Liebig formulated the law of the minimum: plant growth is limited by the most deficient nutrient, even if the other elements are abundant.

Motile Capable of self-propulsion by means of flagella, cilia, or amoeboid movement.

Mottle Disease symptom comprising light and dark areas in an irregular pattern, usually caused by a virus; often used interchangeably with mosaic.

Mottled leaf 1. A disease caused by a zinc deficiency that reduces the size of leaves and fruits in citrus plants.

2. A disease that causes the leaves of cherry trees to pucker and crinkle.

Movement protein For viruses, a protein encoded by the pathogen that is required for spread within an infected plant.

Mow 1. A pile of hay or sheaves of grain, placed in a barn.

2. To cut down with a sharp tool (i.e. to mow grass with a lawn mower).

mRNA (messenger ribonucleic acid) The message made during transcription by reading the DNA sequence to build a particular protein. A single-stranded nucleic acid similar to DNA but having a ribose sugar rather than deoxyribose sugar and a uracil rather than thymine as one of the bases.

MS (Mass Spectrometry) A powerful analytical technique used to identify and measure a wide variety of biological and chemical compounds. Applications include molecular weight analysis of proteins and oligosaccharides and the identification and characterization of natural products.

MS medium A popular plant tissue culture medium developed by T. Murashige and F. Skoog.

MTOC (microtubule organizing center) mtocs are bundles of protein tubes which may be found at the base of a eukaryotic flagellum. In animals, they also function in creating the arrays of microtubules that pull the chromosomes apart during mitosis.

Mucilage A water-holding substance made of various complex carbohydrates, found in many plant groups.

Mucilaginous Viscous, slimy.

Mucro A short, sharp spur as on a tip of a leaf.

Mucronate Terminated by a sharp tip.

Mucronulate Having a very small mucro.

Mulch Any substance, usually a layer of organic material, applied to the soil over or around a plant to conserve moisture, protect the roots from frost, reduce the growth of weeds, and enrich the soil (i.e. bark chips, rotted manure, garden compost, gravel, black plastic).

Mulching paper A biodegradable sheet made of peat that is used to reduce water evaporation and keep weeds down.

Mule A plant hybrid that is self-sterile and usually cross-sterile due to infertile pollen or undeveloped pistils.

Mull A humus-rich layer of forested soils consisting of mixed organic and mineral matter.

Multicarpellate Having a compound pistil or ovary formed by the joining of several carpels.

Multicellular Of organisms, consisting of more than one cell per organism, and usually exhibiting specialization of different cells for different tasks.

Multigene Family A set of genes that, due to their high degree of sequence similarity, are believed to have evolved from a single ancestral gene.

Multigenic resistance (polygenic resistance) Resistance conferred by several genes .

Multiline Mixture of cultivars that differ only in single, major genes for resistance to a pathogen, giving the opportunity to modify the mixture from year to year as needed.

Multinucleate Having more than one nucleus per cell .

Multiparticulate virus Plant virus in which the genome is divided into more than one piece of nucleic acid, which are encapsidated in more than one particle.

Multipartite virus Plant virus in which the genome is divided into more than one piece of nucleic acid.

Multiple fruit A fruit such as the pineapple, which is actually the fruits of several flowers that have formed in a single cluster.

Multiple-resistance Evolution of populations resistant to herbicides not chemically related and having different sites of action, such as a weed resistant to both sulfonylurea and aryl-.

Multiseptate Having many septa or crosswalls.

Mummy A dried shriveled fruit; plant part or organ partially or completely replaced with fungal structures.

Muriate of potash This is a slow-acting fertilizer that adds oxide of potassium to the soil. Same as potassium chloride.

Muricate Having a rough surface due to the presence of many tiny sharp points.

Musciform Relating to or resembling moss.

Mushroom A fleshy fruiting body of a fungus, especially of a Basidiomycete of the family Agaricaceae.

Mutagen A chemical that induces a high frequency of mutations.

Mutation An alteration in genetic material caused by an actual physical change within the genes of a cell. When such changes occur in the germ cells, they are able to be passed on to the offspring as inherited traits, producing such effects as a leaf shape or flower color different from that of the parent.

Mycelia sterilia A group in the Fungi Imperfecti that does not produce spores, but may produce chlamydospores or sclerotia, e.g. Rhizoctonia and Sclerotium; subsequent research has defined sexual reproduction for some species.

Mycelial Pertaining to mycelium.

Mycelioid Resembling mycelium.

Mycelium The root-like network of filaments (hyphae) making up the nonreproductive part of the body of a fungus.

Mycology The science of fungi; the study of mushrooms.

Mycoparasite A fungus that attacks another fungus.

Mycoplasmalike organism (MLO) Archaic term for phytoplasma; plant-parasitic pleomorphic mollicute (prokaryote with no cell wall) found in phloem tissue; cannot yet be grown on artificial nutrient media.

Mycorrhiza Symbiotic association between a nonpathogenic or weakly pathogenic fungus and the roots of plants .

Mycorrhizae The symbiotic association of the mycelium of a fungus with the roots of certain plants.

Mycotoxin Poisonous compound produced by a fungus.

Mycovirus A virus that infects fungi.

Myosin A protein that interacts with actin as a molecular motor.

Myxamoeba (myxamoebae) Single amoeboid cells of a cellular slime mold (Dictyosteliomycota); these aggregate to form the reproductive structures.

Myxomycetes (slime molds) Saprophytic organisms that form vegetative amoeboid plasmodia and spores.

N

Na+-K+ pump An ion pump that transports Na+ out of the cell and K+ into the cell.

NADPH Reduced nicotinic adenine dinucleotide phosphate

Nanometer n. A unit of measure; one millionth (10-9) of a meter.

Necrosis The death of cells, often accompanied by black or brown darkening of the tissue.

Necrotroph A parasite that typically kills and obtains its energy from dead host cells .

Nectary A nectar-secreting gland in a flower.

Needle cast (of conifers) Disease symptom caused by fungi resulting in premature drop of needles.

Negative cross-resistance Occurs when a resistant biotype is more susceptible to other classes of herbicides than the susceptible biotype.

Negative sense RNA Ribonucleic acid complimentary to the positive or plus sense; not translated into protein .

Nematicide Agent, usually a chemical, that kills nematodes.

Nematode Nonsegmented roundworm (animal), parasitic on plants or animals, or free living in soil or water.

Neotropics The biogeographic region of the New World that stretches southward from the Tropic of Cancer and includes southern Mexico, Central and South America, and the West Indies.

Nernst equation The relationship between ion concentration and membrane potential.

Net Primary Production Accumulation of total biomass over a given period of time. Gross primary production includes plant respiration.

Neurofilament A type of intermediate filament that supports the axons of nerve cells.

Neuron A nerve cell specialized to receive and transmit signals throughout the body.

Neurotransmitter A small, hydrophilic molecule that carries a signal from a stimulated neuron to a target cell at a synapse.

Nexin A protein that links microtubule doublets to each other in the axoneme.

Nicotinamide-adenine dinucleotide (NAD+) A coenzyme that functions as an electron carrier in oxidation/reduction reactions.

Nitrate (NO_3^-) A form of nitrogen that is available to plants as a nutrient. In high concentrations, it is a water pollutant.

Nitrification inhibitor A chemical which, when applied to the soil, slows or temporarily prevents the biological oxidation of ammonium to nitrite and nitrate.

Nitrification Nitrification is the process in which ammonium is converted to nitrite and then nitrate. This process naturally occurs in the environment, where it is carried out by specialized bacteria.

Nitrogen An essential chemical element for biological life.

Nitrogen cycle The cycling of the element nitrogen from gaseous forms to various inorganic forms, such as ammonium, nitrates and nitrites, and organic forms, such as nucleic acids and proteins and back to elemental nitrogen.

Nitrogen Dynamics Movement of nitrogen through an ecosystem.

Nitrogen fixation Process of converting dinitrogen gas to chemically reactive forms, where nitrogen combines with other elements such as oxygen, hydrogen, and cargon.

Nitrogen gas The most abundant form of nitrogen existing as a gas, making up the bulk of the gas in the air (~78%). N2

Nitrogen Mineralization This is the process in which organic forms of nitrogen (e.g., proteins in dead plant material) are converted by microbes to inorganic forms of nitrogen (e.g. ammonium and nitrate).

Nitrogen oxides Primary air pollutants produced from internal combustion engines and other industrial processes.

Nitrogen Saturation Occurs when nitrogen inputs saturate the retention capacity of a forest ecosystem. That is, as more and more nitrogen is deposited on forests (due to increases in nitrogen emissions and corresponding increases in deposition), eventually forests will no longer be able to absorb all the nitrogen. This could lead to increased nitrate leaching (which could pollute streams, lakes and drinking water), soil acidification, and maybe even forest health problems.

Nitrogen-phosphorus-potassium fertilizers Manufactured using ammonium phosphate to which potassium (usually potassium chloride) and water have been added; additional nitrogen and other nutrients also may be added.

Nitrous oxide One of the gaseous form of mineral nitrogen. N2O

NMR (Nuclear Magnetic Resonance) Spectroscopy A non-invasive analytical technique with an infinitely broad range of applications in plant biotechnology and food processing. It is similar to MRI (Magnetic Resonance Imaging) in that it can be used for medical applications. Non-destructive analysis with this tool can be applied to a wide range of research challenges, including the determination of chemical structure, microimaging and plant metabolism studies.

N-myristoylation The addition of myristic acid (a 14-carbon fatty acid) to the N-terminal glycine residue of a polypeptide chain.

NNR National Nature reserve

Node The part of a plant stem from which one or more leaves arise. The nodes are separated on the stem by the internodes.

Nodes A joint on a stem where shoots, leaves and flowers arise

Nodule (nodulation) Small knot or irregular, rounded lump; on leguminous plants, structures on roots that contain nitrogen-fixing bacteria.

Nomadism Continual movement of humans and animals with no fixed settlement, generally in search of food or water

Nomenclature a system of names, or the rules by which a system of names is formed.

Noninfectious disease A disease that is caused by an abiotic agent that cannot be transmitted from one plant to another.

Non-labile phosphorus Slowly available form of phosphorus.

Nonlimiting water range The region bounded by the upper and lower soil water content over which water oxygen and mechanical resistance are not limiting to plant growth .

Nonpathogenic (avirulent) Unable to cause disease.

Nonpersistent transmission (stylet-borne transmission) A type of virus transmission in which the virus is acquired and transmitted by the vector after short feeding times, and is retained by the vector for only a short period of time.

Non-polar A molecule with functional groups lacking dipole moments (e.g. alkyls and aryls)

Non-polar functional groups Functional groups generally containing C and H atoms that do not form dipole moments.

Non-polar solvent A solvent with functional groups lacking dipole moments (e.g. alkyls and aryls)

Nonreceptor protein-tyrosine kinase An intracellular protein-tyrosine kinase.

Non-resupinate Referring to Orchid flowers which don't twist 180 degrees during development or which twist a full 360 degrees, with the result that the labellum is uppermost.

Nonseptate Without cross-walls; coenocytic.

Nontarget genes Genes in the recurrent parent that are not at the locus of the gene of interest. Basically, nontarget genes are all those genes in the recurrent parent that the backcross breeder would like to retain in the backcrossing program.

Normal Conforming with, adhering to, or constituting a norm, standard, pattern, level, or type; typical .

Northern A Method used to transfer RNA molecules separated during electrophoresis to membrane filters for detection of specific nucleotide sequences using probes.

Notifiable disease Diseases of livestock, which must be reported to the police when an outbreak is suspected on a farm. Certain disease and pests must be notified to MAFF. ie Foot & Mouth or Colerado beetle.

No-till A cultural system most often used with annual crops, in which the new crop is seeded or planted directly in a field on which the preceding crop plants were cut down, had the tops harvested, or were destroyed by a nonselective herbicide (the old crop is not removed or incorporated into the soil as is common in preparing a plant bed).

Novel Foods Products that have never been used as a food; foods which result from a process that has not previously been used for food; or, foods that have been modified by genetic manipulation (genetically modified foods).

Nucellar epidermis The outer cellular layer of the nucellus

Nucellar pillar Also called the nucellar projection. A shaft of tissue that runs along the top of the crease on the inner side of the pigment strand

Nucellus The central portion of the ovule, containing the embryo sac, and surrounded by integuments

Nuclear envelope The barrier separating the nucleus from the cytoplasm, composed of an inner and outer membrane, a nuclear lamina, and nuclear pore complexes.

Nuclear export signal An amino acid sequence that targets proteins for transport from the nucleus to the cytosol.

Nuclear lamina A meshwork of lamin filaments providing structural support to the nucleus.

Nuclear localization signal An amino acid sequence that targets proteins for transportation from the cytoplasm to the nucleus.

Nuclear membrane The double membrane which surrounds the eukaryotic nucleus. It has many pores in its surface which regulate the flow of large compounds into and out of the nucleus.

Nuclear membranes Membranes forming the nuclear envelope; the outer nuclear membrane is continuous with the endoplasmic reticulum and the inner nuclear membrane is adjacent to the nuclear lamina.

Nuclear pore complex A large structure forming a transport channel through the nuclear envelope.

Nucleic acid Genetic material of all living organisms, including DNA (deoxyribonucleic acid) and RNA (ribonucleic acid).

Nucleic acid hybridization The formation of double stranded DNA and/or RNA molecules by complementary base pairing.

Nucleoid Region in prokaryotes, chloroplasts, and mitochondria where the DNA is concentrated. Unlike a nucleus, it is not bound by a membrane.

Nucleolus An organelle within the nucleus it is where ribosomal RNA is produced.

Nucleophile Seeks positive charge; may be negatively charged or neutral.

Nucleoside A purine or pyrimidine base linked to a sugar (ribose or deoxyribose).

Nucleosome The basic structural unit of chromatin consisting of DNA wrapped around a histone core.

Nucleosome remodeling factors Proteins that disrupt chromatin structure, allowing transcription factors to bind nucleosomal DNA.

Nucleotide A subunit of DNA or RNA consisting of a nitrogenous base (adenine, guanine, thymine, or cytosine in DNA; adenine, guanine, uracil, or cytosine in RNA), a phosphate molecule, and a sugar molecule. Thousands of nucleotides are linked together to form a DNA strand.

Nucleotide excision repair A mechanism of DNA repair in which oligonucleotides containing damaged bases are removed from a DNA molecule.

Nucleus Spherical body containing many organelles, including the nucleolus. The nucleus controls many of the functions of the cell (by controlling protein synthesis) and contains DNA (in chromosomes). The nucleus is surrounded by the nuclear membrane.

Nutraceutical A product isolated or purified from food that is generally sold in medicinal forms not usually associated with food. A nutraceutical has been demonstrated to have a physiological benefit or provide protection against chronic disease.

Nutrient cycle Pathway of a nutrient through an ecosystem from assimilation (transformation into living tissue) by organisms to release by decomposition.

Nutrient removal Concept whereby fertilizer recommendations are based on how much nutrient a specific crop would be likely to remove from the soil and how much nutrient would be needed to replace it; developed before reliable soil testing was readily available; does not account of the soils' ability to supply many essential nutrients.

O

O Line A structure in young embryo which is the lower boundary of shoot apical tissues.

Obcordate Inversely heart-shaped. The wider end and its deep notch at the top instead of the bottom.

Obdeltoid Inversely deltoid. Triangular, with the apex downward.

Obdiplostemonous Having the stamens in two whorls, those of the outer whorl being opposite the petals.

Obdiplostemony The condition of being obdiplopstemonous.

Oblanceolate Inversely lanceolate; shaped like a lance-point reversed, with the tapering point attached to the leafstalk.

Obligate Restricted to a particular way of life.

Obligate Parasite An organism that is only capable of living as a parasite in association with its host plant. The term is synonymous with biotroph.

Oblique Slanting with unequal sides.

Oblong Longer than broad, and mostly parallel-sided, such as an oblong leaf.

Obovate Inversely ovate; having the broad end upward or toward the apex, as in many leaves.

Obrotund Approaching a round form.

Obtuse Blunt or rounded at the extremity, as an obtuse leaf, sepal, or petal.

Occlusion Block or plug that stops flow of liquids (as in vessels).

Ocrea A sheathing stipule, or a pair of stipules united into a tubular sheath around the stem.

Odd-pinnate Pinnate leaves (leaves w/ opposite pairs) with a terminal odd leaflet, as in the rose.

Odoriferous Giving odor or scent, usually sweet; diffusing fragrance; fragrant, as odoriferous flowers.

Oedema (also edema; intumescence) Blisters produced on leaves and other plant parts under conditions of high moisture and restricted transpiration.

Offset 1. A young plant produced by the parent, usually as its base; also called offshoot.

2. A small bulb at the base of a mother bulb.

3. To produce offsets.

Offsets A young plant produced at the base of a parent plant

Ography The combination of land slopes and the way those slopes interrelate. ography is one of the conventionally accepted soil forming factors.

Oil additives An additive made either synthetically or extracted from seeds that enhances herbicide uptake into the plant; crop oils, methylated seed oils, crop oil concentrates.

Oilseeds Seeds that are used in the production of cooking oils and other food products, such as margarine. Oilseeds include canola, soybean, flaxseed, safflower and sunflower seeds. (Agriculture and Agri-Food Canada)

Okazaki fragments Short DNA fragments that are joined to form the lagging strand of DNA.

Old-growth forests Forest or woodland having a mature ecosystem characterized by the presence of old woody plants and trees, and the wildlife and smaller plants associated with them.

Oligogenic resistance Resistance conferred by a few genes (see monogenic resistance, multigeneic resistance, polygenic resistance, single gene resistance).

Oligonucleotide Small fragments of nucleotides hooked together; primers used in PCR.

Oligosaccharide A short polymer of only a few sugars.

Oligotrophic Term applied to a body of water low in nutrients and in productivity. Mirror Lake is oligotrophic.

Omnivores Animals that feed on material of both plant and animal origin

Oncogene A gene capable of inducing one or more characteristics of cancer cells.

Ontogeny The developmental history of an individual, from its beginning as a single cell (zygote, meiospore, or asexual spore) or group of cells (gemma, soredium) to its mature multicellular structure.

Oogamous Being reproduced by the fusion of a small, mobile male gamete with a large, immobile female gamete.

Oogonium Female gametangium of oomycetes, containing one or more gametes.

Oomycetes Class of fungal-like organisms typically with nonseptate mycelium, asexual sporangia and zoospores, and sexual oospores.

Oospore Thick-walled, sexually-derived resting spore of oomycetes.

Ooze Mass of bacterial cells mixed with host fluids.

Open reading frame Portion of a nucleic acid molecule that is translated into a protein.

Open-pollinated Pollinated by wind or insects, not by human intervention.

Operator A regulatory sequence of DNA that controls transcription of an operon.

Operculum (opercula) in mosses, the lid of the meiosporangium.

Operon A group of adjacent genes transcribed as a single mRNA.

Opposite Leaves, which are arranged in pairs. Two leaves grow from one node directly across from each other on each side of a stem.

Orbicular Nearly or completely circular, as an orbicular leaf. Also orbiculate.

Orchard 1. A garden.

2. A piece of ground, usually enclosed, devoted to the cultivation of fruit trees, especially apple, pear, peach, plum, and cherry trees. A collection of cultivated fruit trees.

Order A group of soils in the broadest category. In the current system of soil classification in the U.S., there are 12 orders, differentiated by the presence or absence of diagnostic horizons: Alfisols, Andisols, Aridisols, Entisols, Gelisols, Histosols, Inceptisols, Mollisols, Oxisols, Spodosols, Ultisols, Vertisols.

Ordinary superphosphate (0-20-0 + 12S) Produced by reacting sulfuric acid with finely ground rock phosphate; about 85-90 percent water soluble; excellent source of phosphorus and sulfur.

Organelle A membrane-bound structure within a cell having a specialized function, e.g. mitochondria and chloroplasts.

Organic chemical A chemical that is created naturally by using an organism(s) from the environment.

Organic Compounds containing carbon derived from decomposed animal or plant organisms. Is also used loosely of mulches, soil mix, etc. taken from plant materials.

Organic compounds Molecule that contains carbon atoms, usually combined with each other and with atoms of one or more other elements.

Organic Farming Basically farming without the use of manufactured chemicals, but now conforming to very detailed production standards

Organic gardening A method of gardening using only organic materials, such as manure, peat moss, and compost, and rejecting inorganic material, such as rock phosphate, ammonium, insecticides, and pesticides.

Organic matter Plant or animal matter, either living or dead. Organic matter is largely made up of organic carbon, oxygen, hydrogen, and nitrogen.

Organic matter The total of the organic compounds in soil, exclusive of undecayed plant and animal tissues, their "partial decomposition" products, and the soil biomass. The term is often used synonymously with humus.

Organic molecule A molecule containing carbon atoms.

Origin of replication A specific DNA sequence that serves as a binding site for proteins that initiate replication.

Origin replication complex (ORC) A protein complex that initiates DNA replication at yeast origins.

Ornithologist The branch of zoology that deals with the study of birds.

Orthostichous Exhibiting orthostichy; arranged in vertical ranks.

Orthostichy A vertical rank; an arrangement of leaves or scales at different heights on an axis so that their median planes coincide, as the vertical ranks of leaves on a stem.

Orthotropic Of or pertaining to or exhibiting orthotropism; growing more or less vertically.

Orthotropism Vertical growth; the tendency of plants to grow more or less vertically.

Orthotropous A term describing a condition in which all parts of a seed grow in a straight line.

Osequence A sequence of related soils that differ, one from the other, primarily because of ography as a soil-formation factor.

Osmophore A gland, usually floral, which produces fragrances that attract pollinators.

Osmoprotection A survival strategy brought about by the synthesis of groups of proteins (termed dehydrins) in response to water stress. Most tissues produce these proteins which protect cellular organelles and membranes from the damaging effects of water loss (desiccation). A notable exception being the cells of the starchy endosperm which undergo programmed cell death.

Osmosis The diffusion of fluids through a semipermeable membrane, as in a living cell.

Osmoticum Particles that cause osmosis.

Ostiole Pore; opening in the papilla or neck of a perithecium, pseudothecium, or pycnidium through which spores are released.

Outcrossing The transfer of genetic information among plants, generally through pollen. Outcrossing occurs for all plants whether genetically modified or not.

Outerbark Exterior, nonliving portion of bark in woody plants.

Ovary The enlarged, rounded, ovule-producing base of the female portion (the pistil) of a flower that eventually develops into a fruit after being fertilized.

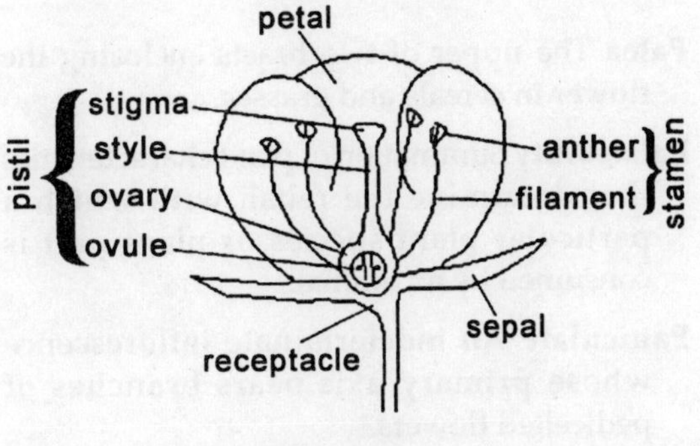

Fig. Ovary

Ovate (of leaves) The leaf is egg-shaped; the broadest part is below the middle.

Overseason To survive or persist from one planting season to the next.

Overwinter To survive or persist through the winter period.

Oviposit To deposit or lay eggs with an ovipositor.

Ovule The part of the plant that, after fertilisation, develops into the seed. It is composed of the embryo sac, the nucellus, the integuments and the funiculus.

Ovum Unfertilised egg-cell containing a haploid nucleus.

Oxic Oxygen present.

Oxidation Removal or exchange of electorns usually by combining with oxygen.

Oxidative phosphorylation The synthesis of ATP from ADP coupled to the energetically favorable transfer of electrons to molecular oxygen as the final acceptor in an electron transport chain.

Ozone A secondary air pollutant that is a highly reactive form of oxygen (O3) formed when the exhaust of automobiles and other internal combustion engines is in the presence of sunlight.

Ozone layer A layer of ozone found in the stratosphere, where it absorbs harmful ultra violet radiation

P

p53 A transcription factor (encoded by the p53 tumor suppressor gene) that arrests the cell cycle in G1 in response to damaged DNA and is required for apoptosis induced by a variety of stimuli.

P-680 The reaction center of photosystem II which best absorbs light at a 680 nm wavelength. P-680 is a chlorophyll a molecule.

P-700 The reaction center of photosystem I which best absorbs light at a 700 nm wavelength. P-700 is a chlorophyll a molecule.

Pachytene The stage of meiosis I during which recombination takes place between homologous chromosomes.

Palmitoylation The addition of palmitic acid (a 16-carbon fatty acid) to cysteine residues of a polypeptide chain.

Palisade parenchyma The tissue found beneath the upper epidermis of leaves, composed of elongate, tubular cells aligned perpendicular to the leaf surface. Palisade parenchyma cells generally contain an abundance of chloroplasts.

Paleoecology The branch of ecology that deals with the interaction between ancient organisms and their environment.

Palea The upper of two bracts enclosing the flower in cereals and grasses.

Palatability Summation of plant characteristics that determine the relish with which a particular plant species or plant part is consumed by an animal.

Paniculate An indeterminate inflorescence whose primary axis bears branches of pedicelled flowers.

Panicle An indeterminate inflorescence whose primary axis bears branches of pedicelled flowers.

Pandemic A widespread and destructive outbreak of disease simultaneously in several countries.

PAN (peroxyacetyl nitrates) A secondary air pollutant formed when the exhaust of automobiles and other internal combustion engines is in the presence of sunlight.

Papilla A hump or swelling. In plant pathology, papilla refers to the accumulation of material between the host plant's cell wall and cell membrane at the point of fungal penetration. A papilla is usually composed of silicon, lignin and proteins.

Paraphyletic group a group consisting of a common ancestor and some, but not all, of its descendants.

Paramylon The storage polysaccharide of the Euglenophyta.

Parallel venation Arrangement of leaf vascular bundles in parallel along the long axis of a leaf; characteristic of monocot flowering plants.

Paragynous Having the antheridium contact the oogonium on the side, as in many Pythium sp.

Paracrine signaling Local cell-cell signaling in which a molecule released by one cell acts on a neighboring target cell.

Paraphysis (paraphyses) A hair-like structure associated with gametangia or sporangia; present in mosses, rockweeds, and some other groups.

Parasexualism A type of recombination, found in certain heterokaryotic fungi, that is based on mitosis rather than meiosis. Genetically distinct haploid nuclei fuse in the heterokaryon. The resulting diploid nuclei multiply by mitotic division, with some crossing-over, and a diploid homokaryon develops. During repeated mitotic divisions, chromosomes may be shed by the diploid nucleus to produce a haploid nucleus.

Parasite An organism that derives its nourishment from another organism, with which it lives in close association and which is harmed by the association.

Parenchyma The soft tissue comprised of living, thin-walled cells of variable size and form. Parenchyma cells are the most abundant cell type in plants.

In leaves, parenchyma is differentiated into two forms: - Palisade parenchyma: the tissue found beneath the upper epidermis of leaves, composed of elongate, tubular cells aligned perpendicular to the leaf surface. Palisade parenchyma cells generally contain an abundance of chloroplasts. - Spongy parenchyma: the tissue typically found between the palisade parenchyma and the lower epidermis of leaves. Spongy parenchyma consists of loosely arranged, irregularly shaped cells containing chloroplasts and is interspersed with intercellular spaces.

Parenchyma cell A type of plant cell responsible for most metabolic activities.

Parent material Rock or minerals which are weathered to form smaller particles of a soil. Parent material is one of the five factors contributing to formation of a specific soil. In the Great Plains, much parent material is associated with ancient seas or glacial deposits.

Parsimony The principle that the simplest explanation, the one that requires the fewest hypotheses, is the one most likely to be correct.

Parthenogenesis Reproduction by the development of an unfertilized egg.

Partial resistance Resistance in which disease develops more slowly on the host or to a lesser extent, but the host does become diseased .

Particle density The weight of an individual soil particle per unit volume. Particle density is usually expressed in units of grams per cubic centimeter (g/cm2). Bulk density considers both the solids and the pore space; whereas, particle density considers only the mineral solids.

Passive diffusion The diffusion of small hydrophobic molecules through a phospholipid bilayer.

Passive transport The transport of molecules across a membrane in the energetically favorable direction.

Pasteurization To free a material, usually a liquid, of selected harmful microorganisms using heat.

PAT The enzyme that controls resistance of Liberty herbicide by detoxifying the herbicide molecule.

Patchy distribution Close and often repeated grazing of small patches or even individual plants while adjacent similar patches or individual plants are lightly grazed or not grazed.

Pathogen An organism, usually a microbe, that causes adverse effects in its host that may cause disease.

Pathogenesis The sequence of processes in disease development that describes a pathogen's association with its host. The sequence begins with initial contact between the pathogen and host and ends when the pathogen is no longer associated with that host (i.e. when the host/pathogen dies or the pathogen moves to another host).

Pathogenesis-related (PR) proteins Proteins, such as antimicrobial proteins and hydrolytic enzymes, that are synthesized in the early events of the plant defense response.

Pathogenicity The ability to cause disease.

Pathology The study of diseases.

Pathotype A subdivision of a pathogen species characterized by its pattern of virulence or avirulence to a series of differential host varieties.

Pathovar (abbr. pv.) A subdivision of a plant pathogenic bacterial species defined by host range; pathovar for bacteria is equivalent to forma specialis for fungi.

Pathway A sequence of interconnecting enzyme reactions.

PCR (polymerase chain reaction) A method for replicating a particular sequence of DNA in vitro. Used to generate greater amounts of DNA for analysis or to determine if a particular sequence exists.

Pectin A methylated polymer of galacturonic acid found in the middle lamella and the primary cell wall of plants; jelly-forming substance found in fruit.

Ped A unit of soil structure, formed together by joining fine sand, silt, or clay particles

Pedicel Small slender stalk; stalk bearing an individual flower, inflorescence, or spore.

Peduncle Stalk or main stem of an inflorescence; part of an inflorescence, or a fructification.

Peduncle The final internode of the plant stem, that bears the flower.

Pellicle A layer of elastic proteins beneath the plasmalemma of Euglenophyta.

Penetration Initial invasion of a host by a pathogen.

Penetration Peg A structure found in some plant parasitic fungi. The penetration peg is a specialised, narrow, hyphal strand located on the underside of an appressorium that penetrates the epidermal cell wall.

Penicillin Antibiotic produced by certain species of Penicillium and effective against Gram-positive bacteria.

PEP carboxylase promoter Phosphenol pyruvate (PEP) carboxylase is an enzyme used in photosynthesis. The promoter of this gene induces protein production in actively photosynthesizing tissues of plants.

Peptide bond The bond joining amino acids in polypeptide chains.

Peptidoglycan The principal component of bacterial cell walls consisting of linear polysaccharide chains crosslinked by short peptides.

Peptidyl prolyl isomerase An enzyme that facilitates protein folding by catalyzing the cis-trans isomerization of prolyl peptide bonds.

Percolation The downward movement of water through the soil, made possible by pore space in the soil.

Perennial Something that occurs year after year; plant that survives for several to many years .

Perfect A flower that has both functioning pistils and stamens

Perfect flower A flower with both stamens and pistils. Sepals and petals may or may not be present.

Perfect state (teleomorph) The sexual form in the life cycle of a fungus .

Perianth Collectively, the calyx and the corolla.

Pericarp The ovary wall; later the fruit/seed/grain wall.

Pericentriolar material The material in the centrosome that initiates microtubule assembly.

Pericycle Layer or layers of cells between the phloem and the endodermis of roots, giving rise to branch roots.

Peripheral membrane proteins Proteins indirectly associated with cell membranes by protein-protein interactions.

Periphyton Sessile organisms, such as algae and small crustaceans, that live attached to surfaces projecting from the bottom of a freshwater aquatic environment.

Peristome A ring of inward-directed teeth surrounding the opening of a moss capsule. The peristome teeth move in response to humidity, thus facilitating spore dispersal.

Perithecium (perithecia) Flask-shaped or subglobose, thin-walled fungus fruiting body (ascocarp) containing asci and ascospores; spores are expelled or released through a pore (ostiole) at the apex.

Peritrichous Having hairs or flagella distributed over the whole surface.

Permanent pasture An established plant community in which the dominant species are perennial grasses, there are few or no shrubs and trees are absent

Permanent wilting point The soil water content when a plant has extracted all the water it can. At this point, a plant will wilt and not recover.

Permineralization Slow replacement of most or all of the organic material in a fossil with mineral crystals (usually silicate).

Peroxisome A cytoplasmic organelle specialized for carrying out oxidative reactions.

Peroxyacetal nitrates (PAN) A secondary air pollutant formed when the exhaust of automobiles and other internal combustion engines is in the presence of sunlight.

Persistent transmission (circulative transmission) A type of virus transmission in which the virus is acquired and transmitted by the vector after relatively long feeding times and remains transmissible for a prolonged period while in association with its vector.

Pest Any organism that damages plants or plant products.

Pesticide Any substance or mixture of substances used for controlling, preventing, destroying, repelling, or mitigating any pest. A pesticide also includes any substance or mixture of substances intended for use as a plant regulator, defoliant, or dessicant.

Petal A modified leaf, usually colored, forming a part of the corolla of a flower; occurs within the sepals (calyx).

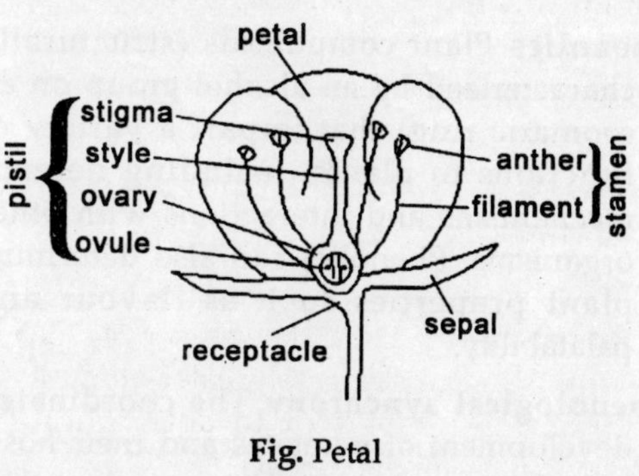

Fig. Petal

Petiole The basal cylindrical stalk of a leaf.

PGPRs (Plant Growth Promoting Rhizobacteria) A class of bacteria that enhance canola plant growth by preventing deleterious effects of other microorganisms that are harmful to plants.

pH Negative logarithm of the effective hydrogen ion concentration; a measure of acidity (pH 7 is neutral; values less than pH 7, acidic; greater than pH 7, alkaline).

pH scale A scale from 0 to 14 reflecting the concentration of hydrogen ions; the lower numbers denote acidic conditions, 7 is neutral, and the higher numbers denote basic (alkaline) conditions.

Phagocytosis The uptake of large particles, such as bacteria, by a cell.

Phalloidin A drug that binds to actin filaments and prevents their disassembly.

Phellem Cork; a protective tissue composed of nonliving cells with suberized walls produced by the phellogen and accumulating on the surface of stems and roots.

Phelloderm Tissue formed by and internal to the phellogen; resembles the cortical parenchyma in morphology.

Phellogen Cork cambium; lateral meristem forming the periderm, a protective tissue in stems and roots; phellem (cork) is produced toward the surface, phelloderm toward the inside.

Phenolics Plant compounds (structurally characterized by an alcohol group on an aromatic ring) that impart a variety of functions to plants, including defence mechanisms and interactions with other organisms. Phenolics can also determine plant properties such as flavour and palatability.

Phenological synchrony The coordinated development of parasites and their hosts based on degree-days and other environmental factors so that parasites can successfully infect their hosts, such as the production and release of ascospores of Venturia inaequalis, the apple scab pathogen, from fallen leaves from the previous season, when new apple leaves are expanding.

Phenology Study of the timing of biological activity over the course of a year, particularly in relation to climate.

Phenotype The observable physical characteristics of an organism that are determined by a combination of the genetic composition (genotype) and the environment of the individual.

Phenotypic trait An observable physical characteristic of an organism.

Phenoxyacetic acids A family of herbicides with similar chemical structure belonging to the growth regulating mode of action. An example would be 2,4-D.

Pheromone A chemical secreted by an insect or animal that influences the behavior of others within the same species; often functions as an attractant of the opposite sex.

Phialide End cell of a conidiophore with one or more open ends through which a basipetal succession of conidia develops.

Phloem Tissue used for the movement of sugars from source tissue to sink tissue; Composed of living cells called sieve elements that have no nuclei, vacuoles, or ribosomes. The ends of the sieve elements contain connective pores, which join them together to form tube-like structures.

Phloem necrosis Death of phloem cells, often visible, caused by infection by systemic phloem pathogens such as phytoplasmas, spiroplasmas, and viruses .

Phohosphorylation The coupling of photosystems I and II with an electron transfer chain that moves electrons from water (which is oxidized to form O_2) to NADP+ (which is reduced to form NADPH). The transfer of electrons between photosystem II and photosystem I releases energy, which is conserved in the form of a trans-membrane proton gradient and used to synthesize ATP.

pHopt The optimum pH for a reaction.

Phorbol esters A class of tumor promoters that stimulate protein kinase C by acting as analogs of diacylglycerol.

Phosphatase An enzyme that some algae species produce under conditions of low P availability to convert organic P to inorganic P that the algae can utilize for growth.

Phosphate Used, rather than phosphorus, in fertilizer technology to refer to available phosphoric acid (P2O5).

Phosphatidylinositide 3-kinase (PI 3-kinase) An enzyme that phosphorylates PIP2, yielding the second messenger phosphatidylinositol 3,4,5-trisphosphate (PIP3).

Phosphatidylinositol 4,5-bisphosphate (PIP2) A minor phospholipid component of the inner leaflet of the plasma membrane. Hormones and growth factors stimulate its hydrolysis by phospholipase C, yielding the second messengers diacylglycerol and inositol trisphosphate.

Phosphodiester bond A bond between the 5′-phosphate of one nucleotide and the 3′-hydroxyl of another.

Phosphoinositide Phospholipids containing inositol involved in linking extracellular signals to intracellular responses, or signal transduction.

Phospholipase C An enzyme that hydrolyzes PIP2 to form the second messengers diacylglycerol and inositol trisphosphate.

Phospholipid bilayer The basic structure of biological membranes, in which the hydrophobic tails of phospholipids are buried in the interior of the membrane and their polar head groups are exposed to the aqueous solution on either side.

Phospholipid molecules A lipid that is mostly nonpolar, but contains a charged phosphate group and possibly other polar functional groups at one end. This type of lipid is a primary component of biological membranes

Phospholipid transfer protein A protein that transports phospholipid molecules between cell membranes.

Phosphorescence Light emitted by an excited triplet molecule as part of the de-excitation process. Because a significant amount of the energy in the original exciting photon is lost in the conversion from the excited singlet to the excited triplet state, the wavelength of phosphorescence is greater (i.e. less energy) than the wavelength of the exciting photon and greater than the wavelength of any fluorescence from the excited singlet state in that molecule.

Phosphoric acid Produced by treating finely ground rock phosphate with sulfuric acid or by smelting (burning) rock phosphate with coke and silica in an electric furnace and reacting with water. Agricultural grade phosphoric acid is seldom used for direct application because of its corrosive properties; it is usually used to make other forms of phosphoric fertilizers.

Phosphorus Next to nitrogen, the most limiting nutrient in Nebraska crop production; naturally found in sufficient amounts in many Nebraska soils; component of the complex nucleic acid structure of plants which regulates protein synthesis; important in cell division and development of new tissues.

Phosphorus, bio-available This water P fraction, also refer to algal-available P, is a combination of dissolved P plus the estimated portion of P to be released from sediment P within a short time, e.g. 45 hours, after the particles enter surface water.

Phosphorus, dissolved This water P fraction, a near equivalent with dissolved reactive P, is readily available to aquatic plants and animals.

Phosphorus, labile This soil P fraction is typically accounts for less than 1% to less than 5% of the P in soils, and is less tightly bond than stable P. Some labile P, as well as solution P, is measured with agronomic soil tests.

Phosphorus, sediment This water P fraction, also referred to as particulate P, often is the greatest fraction in water bodies. It consists of organic and inorganic sediments. Its not available to most aquatic plants and animals, but phosphatase producing algae can cause organic sediment P to become available.

Phosphorus, solution This soil P fraction is typically accounts for less than 1% of the P in soils and is readily available to plants.

Phosphorus, stable This soil P fraction is also called non-labile P. It accounts for most of the P in soils, typically more than 95% of the total, and includes tightly bond P in secondary and primary minerals and in organic forms.

Phosphorylation The addition of a phosphate group to a molecule.

Phosphotidylinositol Is a membrane lipid called a phosphoinositide. Phosphoinositides are phospholipids containing inositol involved in linking extracellular signals to intracellular responses, or signal transduction.

Photochemical Chemical reactions that require sunlight.

Photomorphogenesis The development of a plant cell or organ that is triggered by exposure to light. For example, if a corn plant is germinated in the dark, it will produce an etiolated shoot, lacking chlorophyll or the ability to carry out photosynthesis. Exposing an etiolated shoot to light will initiate development of pigment biosynthesis and the production of cellular components necessary for photosynthesis.

Photon The basic unit of light. Although light has some properties consistent with a wave, light from the sun is in individual packages. Each photon is a particle of electromagnetic radiation traveling with the speed of light (3 X 108 m sec-1). The energy in each photon is dependent on the frequency with which the electromagnetic field vibrates.

Photoperiodism The effect of day and night length on plant flowering. Some plants are long day, requiring14 16 hours of sunlight per day to flower. Others are short day requiring only 8 9 hours. Others are day neutral and unaffected by day length. Photoperiosim also effects the breeding cycles of farm livestock

Photophosphorylation The coupling of photosystems I and II with an electron transfer chain that moves electrons from water (which is oxidized to form O_2) to NADP+ (which is reduced to form NADPH). The transfer of electrons between photosystem II and photosystem I releases energy, which is conserved in the form of a trans-membrane proton gradient and used to synthesize ATP.

Photoreactivation A mechanism of DNA repair in which solar energy is used to split pyrimidine dimers.

Photosensitizers Molecules which have efficient conversions to the triplet state after absorption of a photon. These triplet state molecules either reduce other molecules to generate free radicals, or they excite O_2, normally a triplet, to the excited singlet state; singlet oxygen is very reactive and can cause significant cellular damage.

Photosynthate Product of photosynthesis.

Photosynthesis A process in which plants convert sunlight, water, and carbon dioxide into food energy (sugars and starches), oxygen and water. Chlorophyll or closely-related pigments (substances that color the plant) are essential to the photosynthetic process.

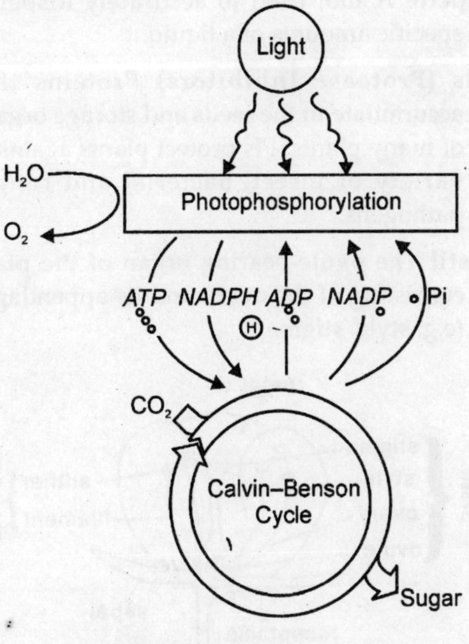

Fig. Photosynthesis

Photosynthetic inhibitors The mode of action of a group of herbicides that directly influence the light reaction of photosynthesis.

Photosystem An array of pigment-protein complexes and electron transfer components that function together to harvest light energy, transfer the energy to photochemical reaction centers, and move the excited electrons in a controlled fashion to produce usable biochemical energy. Each photosystem contains hundreds of chlorophyll and carotenoid molecules functioning as antennae, while only a few chlorophyll molecules are employed in the reaction centers.

Phycobilin A water-soluble photosynthetic accessory pigment of the Rhodophyta and cyanobacteria.

Phycomycete Archaic term for a member of a group of fungi without cross walls (septa) in their mycelium.

Phyllody Change of floral organs to leaflike structures.

Phylloplane Leaf surface.

Phylloplane competent Used to describe epiphytic microorganisms adapted to living on the leaf surface.

Phyllotaxy Arrangement of leaves on a stem in relation to one another.

Phylogeny Genealogy of species; the pattern of ancestry and descent of species in evolution.

Physical structure The combination or arrangement of soil particles into secondary units or peds.

Physiology The biological study of the functions of living organisms and their parts. Scientists studying plant physiology might be interested in how nutrients are transported from roots to leaves.

Phytoalexin A substance produced in higher plants in response to a number of stimuli (chemical, physical or biological) that inhibits the development of a microorganism.

Phytoanticipin These are constitutive phytoalexins i.e. phytoanticipins are synthesised by the plant at a constant rate and therefore always present in the tissues of the plant, whereas phytoalexins are only produced in response to a stimulus such as a pathogenic invasion. The distinction between phytoalexins and phytoanticipins is not always clear as some compounds may be phytoalexins in one species and phytoanticipins in another species. In general, the distinction between the two compounds depends on when they are produced (either before or after infection) and the extent to which the compound is antimicrobial (phytoalexins are antimicrobial).

Phytohormones Compounds produced by the plant to regulate growth and development of the plant.

Phytopathogenic Able to cause disease in plants.

Phytopathology (plant pathology) The study of plant diseases.

Phytoplankton Small or microscopic photosynthetic organisms that float or drift in great numbers in fresh or salt water near the surface and are the base of the ocean food chain

Phytoplasma (mycoplasmalike organism, MLO) Plant-parasitic pleomorphic mollicute (prokaryote with no cell wall) found in phloem tissue; cannot yet be grown on artificial nutrient media.

Phytoremediation The use of plants to clean up sites that have been contaminated with chemicals or petroleum products.

Phytosanitary certificate Official document that indicates that plant material has been inspected and found to be free of certain pathogens.

Phytosociology The branch of ecology that deals with the characteristics, classification, relationships, and distribution of plant communities.

Phytotoxic Harmful to plants (usually used to describe chemicals).

Pigment A colored compound, such as chlorophyll, in the cells of plants or fungi.

Pigment strand A cylinder of cells that lies between the vascular tissues and the nucellar projection in the crease region.

Pigment synthesis inhibitors The mode of action of a group of herbicides that ss pigment synthesis (not chlorophyll)i these pigments are present to protect chlorophyll from breakdown by sunlight. Without the pigments present, chlorophyll is easily degraded showing bleaching symptoms on plant tissue.

Pinocytosis The uptake of fluids or molecules into a cell by small vesicles.

Pipette A tool used to accurately dispense specific amounts of a liquid.

PIs (Protease Inhibitors) Proteins that accumulate in the seeds and storage organs of many plants. PIs protect plants against a variety of insect, bacterial and fungal pathogens.

Pistil The ovule-bearing organ of the plant consisting of the ovary and its appendages (e.g. style, stigma).

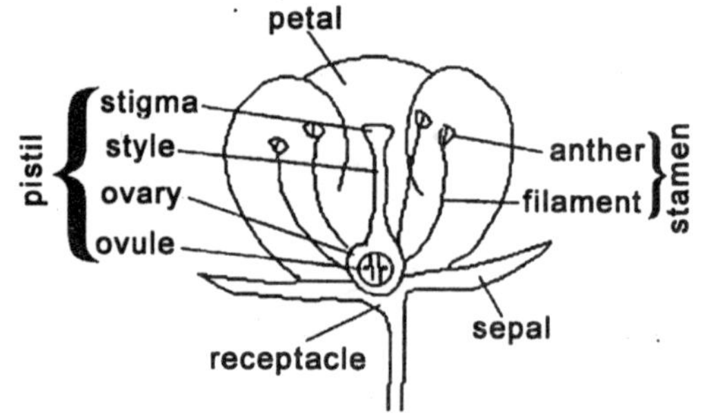

Fig. Pistil

Pith Parenchymatous tissue occupying the center of the stem.

Pith The inner tissue of most stems and some roots.

pKa The dissociation constant of an acid; the pH at which 50% of the molecules are associated (have the H+ ion) 50% are dissociated (lost the H' ion).

Plant available water That portion of the soil water that can be absorbed by a plant. Generally, plant available water is considered to be 50 percent of the water holding capacity.

Plant hormones Plant-produced signal molecules that regulate germination, growth, metabolism and other physiological functions. The major classes include: auxins, cytokinins, ethylene, gibberellins, and abscisic acid (ABA).

Plant pathology (phytopathology) The study of plant diseases.

Plasma membrane Outer membrane of a cell, sometimes called the cell membrane. The term plasma membrane is used more frequently when discussing prokaryotes.

Plasma membrane The semi-permeable membrane that surrounds the protoplasm of a cell

Plasmalemma The membrane that surrounds a protoplast.

Plasmalogens A family of phospholipids that have an ether bond and an ester bond.

Plasmid A small circular piece of DNA from bacteria that often contains antibiotic resistance genes. Some types have the capability to be replicated in the bacteria and integrate itself into the genome of a plant.

Plasmodesma A cytoplasmic connection between adjacent plant cells formed by a continuous region of the plasma membrane.

Plasmodesmata Narrow extensions of cell sap through cell walls

Plasmodiophoromycetes A fungus-like group of organisms, sometimes called the Plasmodiophoromycota or the endoparasitic slime molds, characterized by the production of zoospores and plasmodia that are restricted to the cells of their host.

Plasmodium (plasmodia) The multinucleate mass of protoplasm that constitutes the vegetative body of Myxomycota.

Plasmogamy Fusion of the cytoplasm of two gametes; part of syngamy.

Plasmolysis Shrinking of a protoplast due to water loss from the vacuole of a plant cell.

Plastid Any of several pigmented cytoplasmic organelles found in plant cells and other organisms, having various physiological functions, such as the synthesis and storage of food.

Plate A rigid piece of the Earth's lithosphere that moves horizontally and interacts with other plates along its boundaries.

Platelet derived growth factor (PDGF) A growth factor released by platelets during blood clotting to stimulate the proliferation of fibroblasts.

Plectenchyma A three-dimensional tissue formed of tightly packed filaments.

Pleiotropic Having multiple effects.

Pleiotropy The control of more than one trait by a single gene.

Pleomorphic Able to assume various shapes (and perhaps sizes); occurrence of more than one distinct form in the life cycle of an organism.

Plesiomorphy A character state already present in the ancestral species of a group under study.

Ploidy The number of sets in a cell of basic chromosomes; often affects plant vigor, flower size and number, and ability to hybridize; an important fact in daylilies and orchid species

Plow Operation that completely turns the soil over to a depth of about 8 inches; uses a moldboard plow.

PNTs (Plants with Novel Traits) Any genetic plant strain with characteristics that show neither familiarity nor substantial equivalent traits to plant species already cultivated in Canada.

Poaching Wet soils which are trampled and 'churned up' by cattle, they become muddy and it often leads to a breakdown in soil structure.

Polar At one end or pole of the cell.

Polar axis of plant The longitudinal directions a plant grows.

Polar functional groups Functional groups with atoms such as O, N, S or P which do not equally share the electrons in the bonds, resulting in dipole moments

Polar microtubules Microtubules of the mitotic spindle that overlap in the center of the cell and push the spindle poles apart.

Polar molecules Molecules with functional groups containing dipole moments (e.g. alcohols, carbonyls, amides, amines)

Polar nuclei Two haploid nuclei in the embryo sac. These nuclei fuse with a pollen nucleus to form a triploid endosperm nucleus, which subsequently divides to form the endosperm.

Polar solvent A solvent with molecules containing functional groups with dipole moments

Polarity Property of a covalent bond or an entire molecule that results in a negative pole and a positive pole; A bond becomes polar if the electrons involved in the bond are not shared equally. Polarity results when the center of the negative charge does not coincide with the center of the positive charge.

Polarized light A technique used in light microscopy. A beam of polarized light is passed through thin slices of preserved tissue. The light is diffracted by some parts of the plant tissue and not others. Starch grains diffract the light in a very characteristic manner producing an image of the Maltese Cross. This image is diagnostic of the presence of starch grains.

Pollen Dust-like material produced by the anther (male reproductive structure) that is composed of pollen grains (microgamehytes) which carry the male gametes of seed plants; important in fertilization

Pollen tube The structure formed by germinating pollen that absorbs nutrients; and in some groups carries the sperm cells to the eggs.

Pollination The transfer of pollen from an anther (the male reproductive organ) to a stigma (the receptive part of the female reproductive organ).

Poly-A tail A tract of about 200 adenine nucleotides added to the 3′ ends of eukaryotic mRNAs.

Polyadenylation The process of adding a poly-A tail to a pre-mRNA.

Polycistronic Messenger RNAs that encode multiple polypeptide chains.

Polyclonal antibody A preparation containing antibodies against more than one epitope of an antigen .

Polycyclic Having several to many disease cycles in a growing season .

Polyetic Requiring more than one year to complete one life or disease cycle.

Polygenic Pertaining to, or governed by, many genes .

Polygenic resistance (multigenic resistance) Resistance conferred by several genes (see monogenic resistance, oligogenic resistance, single gene resistance).

Polymerase chain reaction (PCR) A technique used to amplify the number of copies of a specific region of DNA in order to produce enough of the DNA for use in various applications uch as identification and cloning.

Polymorphic Existing in multiple forms. When a marker locus is polymorphic, there are detectable differences in DNA sequence at that locus.

Polynucleotide A polymer containing up to millions of nucleotides.

Polyp A benign tumor projecting from an epithelial surface.

Polypeptide A polymer of amino acids.

Polyphyletic group A group of organisms not including the common ancestor of all members of the group.

Polyploidy Having more than 2n (diploid) number of basic chromosomes such as triploid (3n) or tetraploid (4n); very common in angiosperms

Polyprotein A protein translated from an entire viral genome which is then cleaved by proteases (protein-degrading enzymes) into the active protein products.

Polysaccharide A polymer containing hundreds or thousands of sugars.

Polysome A series of ribosomes translating a messenger RNA.

Pome (pome fruit) Simple, fleshy, indehiscent fruit derived from several carpels, e.g. apple, pear, quince.

Population dynamics The study of factors affecting the variability of populations of plants and animals over time and space.

Pore space Spaces in soil filled with water or air, between the mineral and organic matter.

Porosity Porosity, or pore space, refers to the volume of soil voids that can be filled by water and/or air. It is inversely related to bulk density.

Positive sense RNA RNA that can serve directly as messenger RNA. **Postemergence** Application of a herbicide after the plant has emerged.

Posterior Toward the back or rear;.

Postzygotic mechanisms Barriers to gene flow between to two populations after fertilization causing reproductive isolation that leads to two different species (genotype) of an organism but they may not be distinguishable by appearance (phenotype); barriers include death of zygote, hybrid inviability, hybrid sterility, and hybrid breakdown (reduced viability and fertility of second generation)

Potassium Essential plant nutrient needed in large amounts; vital to plant nutrient absorption, respiration, transpiration, and enzyme activity. The major portion of potassium is contained in minerals such as feldspar and mica, and clays such as montmorillonite, vermiculite and illite.

Potassium chloride Lowest cost and most commonly used potassium fertilizer; mined as sylvinite from underground mines or by solution mining in New Mexico, Utah, California, and Michigan.

Potassium magnesium sulfate Double salt of potassium and magnesium sulfate; commonly used as a dry granular source of potassium, magnesium, and sulfur.

Potassium sulfate Used as a potassium fertilizer for crops where the chloride content of potassium chloride is undesirable.

Powdery Mildew White powdery 'bloom' on the plant surface caused by the production of fungal mycelium, conidiophores and conidia by members of the Erysiphales (powdery mildew fungi).

PPO Proorphyrinogen oxidase (protox); enzyme utilized in photosynthesis and is the target site of many herbicides in the cell membrane disrupter mode of action.

Pre mRNA The primary transcripts that are processed to form messenger RNAs in eukaryotic cells.

Pre rRNA The primary transcript, which is cleaved to form individual ribosomal RNAs (the 28S, 18S, and 5.8S rRNAs of eukaryotic cells).

Pre tRNA The primary transcript, which is cleaved to form transfer RNAs.

Precipitation Any form of water, such as rain, snow, sleet, or hail, that falls to the earth's surface.

Predispose (predisposition) To make prone to infection and disease.

Preference Selection of certain plants, or plant parts, over others by grazing animals.

Preference Selection of certain plants, or plant parts, over others by grazing animals.

Premia payment paid on the number of ewes qualifying for the sheep quota and calculated on the basis of the actual market price set against the EC price guide.

Prenylation The addition of specific types of lipids (prenyl groups) to C terminal cysteine residues of a polypeptide chain.

Prezygotic mechanisms Barriers to gene flow between two populations berfore fertilization causing reproductive isolation leading to two different species (genotype) of an organism but they may not be distinguishable by appearance (phenotype); barriers include geographical isolation, egolocial isolation, seasonal isolation, behavioural isolation, mechanical isolation, and gametic incompatibility

Primary Inoculum Inoculum that initiates disease in the field following a dormant stage in its life cycle (called overwintering or oversummering). Compare with secondary inoculum.

Primary leaf First true leaf that emerges on a plant following the cotyledons.

Primary pollutant Air pollutant that is released directly into the atmosphere and is harmful to plants, e.g. SO_2 or NOx.

Primary producers Living things that produce their own food (e.g., bacteria and plants).

Primary production Production by plants: 'primary' in being the first use of solar radiation, the main energy source for biological processes.

Primary root Root that develops directly from the radicle of an embryo rather than from a crown or node.

Primary structure The sequence of amino acids in a polypeptide chain.

Primase An RNA polymerase used to initiate DNA synthesis.

Primer Small fragment of nucleic acid with a free 3'-hydroxyl group necessary for initiation of DNA, and, sometimes, RNA synthesis; often specific fragments chosen for use in polymerase chain reaction (PCR) for rapid identification of pathogens.

Primer annealing step The second step of PCR. In this step the primers will bind to their complementary DNA sequences on the single-stranded DNA template.

Primer dimers Primer-dimers occur when single stranded primer oligonucleotides bind to each other rather than the DNA template.

Primordia A group of cells that represents the initial stages in development of a plant organ. s primordium

Pro herbicide A molecule that doesn't become herbicidally active until a reaction within a plant makes it a herbicide.

Probe A macromolecule such as DNA or RNA that has been labeled with either radioactivity or antibodies and can be detected by an assay. Probes are used to identify target molecules, genes, or gene products.

Procambium The tissue in the central portion of an embryo.

Processed psuedogene A psuedogene that has arisen by reverse transcription of mRNA.

Production The accumulation of energy or biomass.

Productivity The amount of energy trapped in organic matter during a specified interval at a given trophic level

Proembryonal complex A mass of embryogenic cells.

Profile The shape of a slope profile. Profile is the slope viewed in a vertical cross-section.

Progenitor cell A cell that is able to produce embryogenic cells by cell division.

Progeny The offspring of an organism.

Progeny testing A test of the value of a genotype based on the performance of its offspring produced in some definite system of mating.

Progesterone A steroid hormone produced by the ovaries.

Programmed cell death A normal physiological form of cell death characterized by apoptosis.

Pro-herbicide A molecule that doesn't become herbicidally active until a reaction within a plant makes it a herbicide.

Proinhibitin A plant defense compound that is already present before invasion by a pathogen.

Prokaryote An organism without membrane-bound organelles, and with DNA organized in a single naked circular strand, rather than in chromosomes.

Prokaryotic Literally "before the nucleus", the term applies to all bacteria and archaea. Prokaryotic cells have no internal membranes or cytoskeleton. Their DNA is circular, not linear.

Prolasm All the substance of a cell, usually considered not to include the cell wall. The prolasm of a single cell is a prolast.

Prolast A plant cell that has had the cell wall removed.

Prometaphase A transition period between prophase and metaphase during which the microtubules of the mitotic spindle attach to the kinetochores and the chromosomes shuffle until they align in the center of the cell.

Promoter A specific DNA sequence to which RNA polymerase binds and initiates transcription. This region contains information which regulates when and how often the gene is transcribed and ultimately the amount of protein it produces.

Promycelium (promycelia) In rust and smut fungi, a germ tube issuing from the teliospore and bearing the basidiospores.

Pronuclei Two haploid nuclei in a newly fertilized egg.

Proofreading The selective removal of mismatched bases by DNA polymerase.

Proorphyrin An intermediate in the biosynthetic pathways for chlorophyll and heme. This molecule is a strong photosensitizer and causes much cellular damage if it accumulates in tissues exposed to light.

Propagative transmission (circulative propagative transmission) Pathogen transmission characterized by a long period of acquisition of the pathogen (usually a mollicute, e.g. phytoplasma or spiroplasma, and sometimes a virus) by a vector (typically an insect), a latent period before the vector is able to transmit the pathogen, and retention of the pathogen by the vector for a long period because the pathogen reproduces or replicates in the vector.

Propagules Any part of an organism capable of independent growth (e.g., a spore, a mycelial fragment, etc.).

Propanoic herbicides A family of herbicides that have similar chemical structure and belong to the lipid synthesis inhibiting mode of action (fops).

Prophase The beginning phase of mitosis, marked by the appearance of condensed chromosomes and the development of the mitotic spindle.

Prosthetic groups Small molecules bound to proteins.

Proteasome A large protease complex that degrades proteins tagged by ubiquitin.

Protectant Agent, usually a chemical, applied to a plant surface in advance of a pathogen to prevent infection.

Protectant fungicide (contact fungicide) A fungicide that remains on the surface where it is applied; no after-infection activity.

Protection A principle of plant disease control in which a barrier is placed between the susceptible plant and pathogen (e.g. the use of protective chemical dusts or sprays).

Protein A large molecule composed of one or more chains of amino acids in a specific order. Proteins are necessary for the structure, function, and regulation of an organism's cells, tissues, and organs. Each protein has a unique function determined by its shape.

Protein body A membrane-bound granule of storage proteins.

Protein disulfide isomerase An enzyme that catalyzes the formation and breakage of disulfide (S–S) linkages.

Protein kinase C A family of protein-serine/threonine kinases that are activated by diacylglycerol and Ca_2+ and function in intracellular signal transduction.

Protein phosphatase An enzyme that reverses the action of protein kinases by removing phosphate groups from phosphorylated amino acid residues.

Protein synthesis The production of proteins in a cell. Proteins are chains of amino acids linked in the order determined by the genetic code.

Proteinaceous Of or related to protein.

Proteins Polypeptides with a unique amino acid sequence.

Protein-serine/threonine kinase A protein kinase that phosphorylates serine and threonine residues.

Protein-tyrosine kinase A protein kinase that phosphorylates tyrosine residues.

Protein-tyrosine phosphatase An enzyme that removes the phosphate groups from phosphotyrosine residues.

Proteoglycan A protein linked to glycosaminoglycans.

Proteolysis Degradation of polypeptide chains.

Proteomics A study of an organism's proteins, including the molecular structure of the protein. Protein structure often determines the roles that proteins play in plant physiology.

Prothallium (prothallia) The gametophyte of a fern; any small gametophyte.

Proto oncogene A normal cell gene that can be converted into an oncogene.

Protoderm A primary embryonic tissue that is located on the surface of young embryos.

Proton A positively charged atomic particle; a hydrogen ion H+.

Proton motive force (PMF) Created when protons are extruded from the cytosol by electrogenic H+-ATPases, both at the plasma membrane and the tonoplast, a membrane potential and a pH gradient are created at the expense of ATP hydrolysis. This gradient of electrochemical potential for H+ represents stored free energy in the form of the H+ gradient.

Protonated Having a proton attached, usually it is a H+ when considering herbicides.

Protoplasm all the substance of a cell, usually considered not to include the cell wall. The protoplasm of a single cell is a protoplast.

Protoplast A plant cell that has had the cell wall removed.

Protoporphyrin An intermediate in the biosynthetic pathways for chlorophyll and heme. This molecule is a strong photosensitizer and causes much cellular damage if it accumulates in tissues exposed to light.

Protostele The simplest type of stele, consisting of a single column of vascular tissue.

Proximal Nearest to the point of attachment.

Pseudogene A nonfunctional gene copy.

Pseudopodium An actin-based extension of the plasma membrane responsible for phagocytosis and amoeboid movement.

Pseudothecium (pseudothecia) Perithecium-like fruiting body containing asci and ascospores dispersed rather than in an organized hymenium; an ascostroma with a single locule or cavity and containing bitunicate asci.

PTEN A lipid phosphatase that dephosphorylates PIP3 and acts as a tumor suppressor.

Pubescent Covered with hairs; ex. Stachys, Pulsatilla

Puccinia pathway The region through which rust urediniospores move from southern areas through all grain-producing areas of the United States to Canada each season.

Pulse Crops Grain legumes, such as peas, beans, chickpeas and lentils.

Pupa The nonfeeding stage between the larva and adult, organism undergoes complete transformation within a protective cocoon or hardened case.

Pustule Small blisterlike elevation of epidermis formed as spores emerge.

pv. (abbr. for pathovar) A subdivision of a plant pathogenic bacterial species defined by host range; pathovar for bacteria is equivalent to forma specialis for fungi.

Pycnidiospore Spore (conidium) produced in a pycnidium.

pycnidium (pycnidia) Asexual, globose or flask-shaped fruiting body of certain imperfect fungi producing conidia.

pycniospore (spermatium) Haploid, sexually derived spore formed in a pycnium of rust fungi.

Pycnium (spermagonium) Globose or flask-shaped haploid fruiting body of rust fungi bearing receptive hyphae and pycniospores.

Pyramiding The addition, through plant breeding or genetic engineering, of several resistance genes into a single plant cultivar.

Pyrenoid In hornworts and Chlorophyta, a region of the chloroplast involved in starch formation.

Pyrimidine dimer A common form of DNA damage caused by UV light in which adjacent pyrimidines are joined to form a dimer.

Q

QTL Quantitative trait locus. 1. a locus that influences the expression of a quantitative trait.

2. a chromosome region detected by statistical analysis that is significantly associated with variation for a quantitative trait.

Qualitative characters A character in which variation is discontinuous.

Qualitative resistance Resistance reactions that can be placed in distinct categories, usually conferred by one or a few genes.

Quantify To determine or express the quantity of; to count, or measure using an objective, repeatable method.

Quantitative characters A character in which variation is continuous so that classification into discrete categories is not possible.

Quantitative resistance Resistance reactions that have no distinct classes but vary continuously from resistant to susceptible, the result of few to many genes the individual effects of which may be small and difficult to detect.

Quantitative trait locus Quantitative trait locus. 1. A locus that influences the expression of a quantitative trait.

2. A chromosome region detected by statistical analysis that is significantly associated with variation for a quantitative trait.

Quarantine Legal restriction of the transport of plants and/or plant parts in order to prevent the spread of pests and pathogens. In order to accomplish this plants may be held in isolation (i.e. quarantine) for an extended period of time to ensure that they are free of pests and diseases.

Quaternary structure The interactions between polypeptide chains in proteins consisting of more than one polypeptide.

Quenching The process of causing the de-excitation of a molecule from an excited singlet or triplet state back down to the ground state. Quenching of chlorophyll is an important protective role of carotenoids. If a triplet molecule is not rapidly quenched, it can react with oxygen to generate singlet oxygen. Singlet oxygen can cause much cellular damage.

Quiescent Dormant or inactive.

Quorum sensing The ability of bacteria to interact with each other through a variety of mechanisms; allows a population of bacteria to behave more like a multicellular organism.

R

Race Subgroup or biotype within a species or variety, distinguished from other races by virulence, symptom expression, or host range, but not by morphology.

Race non-specific resistance (general resistance, horizontal resistance) Resistance which is effective against all biotypes of the pathogen. (see durable resistance, specific resistance, vertical resistance).

Raceme A simple, indeterminate inflorescence with pedicelled flowers.

Rachidian Of or pertaining to a rachis.

Rachilla The axis in the centre of a spikelet The secondary axis of the flower spike.

Rachis That portion of an inflorescence, above the peduncle, bearing flowers.

Radiate 1. Spreading outward in a wheel shape from the center, as leaves around a stem.

2. Bearing ray flowers, as do many members of the Daisy family, Compositae.

Radical Pertaining to or growing from the root.

Radicate 1. To take root.

2. To cause to take root; to plant deeply and firmly.

Radicicolous Living upon or in roots.

Radicle Part of the plant embryo that develops into the primary root.

Radiculose Producing numerous rootlets.

Radiocarbon dating A method of dating material that contains the element carbon. The method uses isotopes of carbon, and can determine the age of materials as old as 70,000 years.

Radix The root of a plant.

Raf A protein-serine/threonine kinase (encoded by the raf oncogene) that is activated by Ras and leads to activation of MAP kinase.

Rain gauge A device used to quantitatively measure rainfall.

Raised bed A raised area for growing plants, usually enclosed by landscape timbers, in which the growing medium can be accurately determined.

Ramentaceous Covered with, consisting of, or resembling ramentum

Ramentum A thin, brownish, fringed scale that appears on the young shoots or leaves of many ferns.

Ramicaul The leafy stem of some sympodial Orchids, such as those in subtribe Pleurothallidinae; distinguished from the rhizome.

Ramose Branched; having many branches.

Ramuliferous Having many small branches.

Ramulus A small branch or branchlet.

Ran A small GTP-binding protein involved in nuclear import and export.

Range condition The present state of vegetation of a range site in relation to the natural potential plant community for that site.

Rangeland Land on which the native vegetation is predominantly grasses, grass-like plants, forbs, or shrubs.

RAPD Random Amplified Polymorphic DNA. These markers involve PCR amplification of DNA fragments using primers that are random 10-base oligonucleotides. A single RAPD reaction may amplify several to many DNA fragments. RAPDs are dominant markers and, therefore, less informative than codominant markers because heterozygous individuals cannot be identified. A major advantage is that no prior knowledge of an organism's genome is necessary for their use. A common problem with RAPDs has been lack of reproducibility among labs. The technique is described in more detail in S. Grandillo and T.M. Fulton (2002), Approaches to gene mapping. pp. 101-136. In P.M. Gilmartin and C. Bowler (ed.) Molecular Plant Biology Vol. 1. Oxford Univ. Press, Oxford and New York.

Rapeseed A plant widely grown for its industrial oil in the 1940's. In the 1960's breeding efforts led to the removal of two compounds, erucic acid and glucosinolates, changing the plant to an edible oilseed now called canola.

Raphe The part of the supporting stalk of an ovule that is fused to its outer covering. It is usually shaped like a ridge.

Ras A family of small GTP binding proteins (encoded by the ras oncogenes) that couple growth factor receptors to intracellular targets, including the Raf protein-serine/threonine kinase and the MAP kinase pathway.

Ray flower One of the petal-like parts of the flowers of many of the Compositae (Daisy family), such as the Sunflower and Aster.

Rb A transcriptional regulatory protein encoded by a tumor suppressor gene that was identified by the genetic analysis of retinoblastoma.

Reaction center A specialized component of the photosystem that actually carries out the photochemistry with the excitation energy harvested by antenna pigments.

Receptacle The terminal part of the pedicel to which the flower parts (sepals, petals, stamens, pistils) are attached.

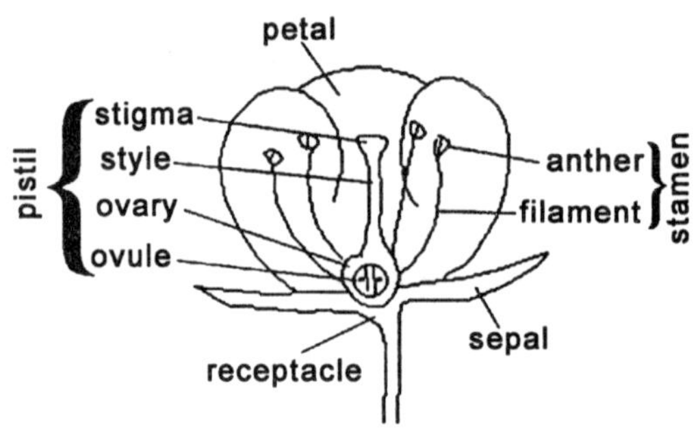

Fig. Receptacle

Receptive hypha The part of a rust fungus pycnium (spermogonium) that receives the nucleus of a pycniospore (spermatium).

Receptor Special proteins in the cell membrane that bind to a signal or chemical messenger from the environment and transmit signals to the cell to initiate a response

Receptor protein-tyrosine kinase Membrane-spanning protein-tyrosine kinases that are receptors for extracellular ligands.

Receptor-mediated endocytosis The selective uptake of macromolecules that bind to cell surface receptors that concentrate in clathrin-coated pits.

Recessive Describes a phenotypic trait that is expressed in diploid organisms only if both parents contribute the trait to the progeny .

Recessive allele An allele whose phenotype is not expressed in a heterozygote.

Recombinant DNA DNA molecules in which sequences, not normally contiguous, have been placed next to each other by in vitro methods.

Recombinant DNA library A collection of genomic or cDNA clones.

Recombinant molecule A DNA insert joined to a vector.

Recombinant plasmid A plasmid that has been cut with special enzymes and 'recombined' in a new order or with DNA from another source.

Recombination Formation of new combinations of genes as a result of the sexual process.

Recombination frequencies/map distance Recombination frequency is the frequency of crossovers between two genes. The map distance (in this case a genetic or recombinational map) is the recombination frequency between two genes times 100. For example, if the frequency of recombination is 0.10, the map distance is 10 map units.

Recurrent parent The parent to which successive backcrosses are made in backcross breeding.

Reduction The opposite of oxidation. When a molecule acquires one or two electrons it becomes more reduced.

Reference Similar to "control," referring to a treatment that reproduces many of the aspects of an experimental design, while excluding the variable of interest. Whereas in an experiment using a true control, the treatment and control differ only in terms of the variable of interest, a reference and a treatment are designed to be as similar as possible, but may have several differences.

Regeneration Regrowth of destroyed parts. Forest regeneration is the process of regrowth after a large disturbance (e.g., by a hurricane or logging operation).

Regiospecificity The ability of an enzyme to target a particular bond with in a molecule.

Regular flower A flower with radial symmetry.

Regulatory protein A protein that binds to the promoter of a gene to regulate transcription.

Regulatory Sequences DNA sequences which direct transcription to occur in the desired tissues and at the correct stages of plant development.

Relative fitness The fitness of one genotype when compared with another.

Relative humidity Water vapour in air compared with the amount of water vapour held at the same temperature when saturated

Remote sensing Way of getting information about an object without having the measuring device in direct contact with it. Examples include radarís use of radio waves and sonarís use of sound waves to detect objects.

Reniform Kidney-shaped.

Repeat sequences The length of a nucleotide sequence that is repeated in a tandem cluster.

Replicase An enzyme involved in the replication (copying) of nucleic acid.

Replicate The copying of a DNA molecule in a cell.

Replicated Copied DNA. To make copies of DNA.

Replication The process by which a DNA or RNA molecule makes an exact copy of itself;

Replication fork The region of DNA synthesis where the parental strands separate and two new daughter strands elongate.

Repressor A regulatory molecule that blocks transcription.

Reproductive success The number of progeny born, or surviving progeny produced by an organism. An animal that has few offspring relative to other animals in its population has low reproductive success, while an animal that has many offspring has a higher reproductive success.

Residence time The ratio of the size of a compartment to the flux through it, expressed in units of time; thus, the average time spent by energy of a substance in the compartment. For example, scientists are often interested in the residence time of certain nutrients in a lake. The lake is the compartment, and the nutrient is the substance. Residence time is a measure of how long this nutrient (e.g. after it arrives via a stream, or deposition, etc.) stays in the lake before being removed (e.g., by the outflow stream).

Resin Sticky to brittle plant product derived from essential oils; often a defense compound of conifers.

Resin duct In seed plants, a tube-like extracellular space lined with resin-producing cells and filled with resin.

Resinosis A disease symptom in which there is a copious flow of resin from a wound or infection site of a conifer.

Resistance The inherited ability of a population to survive and reproduce following repeated exposure to a dose of herbicide normally lethal to the wild type. Resistance may also be induced by such techniques as genetic engineering or selection of variants produced by tissue culture or mutagenesis.

Resistant the inherited ability of a plant to survive and reproduce following exposure to a dose of herbicide normally lethal to the wild type. In a plant, resistance may be naturally occurring or induced by such techniques as genetic engineering or selection of variants produced by tissue culture or mutagenesis.

Respiration An intracellular process which consists of a series of chemical reactions that make energy available through the oxidation of carbohydrates, fats and proteins. This process produces carbon dioxide and water as byproducts and is termed aerobic respiration because it uses oxygen (although the initial stages of the process are anaerobic, i.e. do not require oxygen).

Resting Spore A thick-walled spore, usually formed by a sexual process, that germinates after remaining in a dormant state for an extended period of time.

Restriction endonuclease Enzyme that cleaves DNA at a particular base sequence; sometimes informally referred to as a restriction enzyme.

Restriction Enzymes Enzymes that recognize specific DNA sequences and cleave them to produce DNA fragments of varying length. The ability to specifically cleave DNA sequences at known sites underlies many of the advances made over the last several decades in molecular biology.

Restriction map The locations of restriction endonuclease cleavage sites on a DNA molecule.

Restriction point A regulatory point in animal cell cycles that occurs late in G1. After this point, a cell is committed to entering S and undergoing one cell division cycle.

Reticulate Having netlike markings.

Reticulate venation Arrangement of leaf vascular bundles in a small network; characteristic of some Gnetophyta and dicot flowering plants.

Reticulopodia Long thread-like pseudopodia that branch apart and rejoin, forming a fine network. They are characteristic of forams.

Retroposon A transposable element that moves via reverse transcription of an RNA intermediate.

Retrovirus A virus that replicates by making a DNA copy of its RNA genome by reverse transcription.

Reverse transcriptase Enzyme used to make complementary DNA from a piece of RNA, such as a plant virus.

RFLP Restriction Fragment Length Polymorphism. DNA is digested with a restriction enzyme and the fragments are separated by electrophoresis. The DNA fragments are then transferred to a membrane and hybridized with a labeled DNA probe. Polymorphisms are detected as labeled DNA fragments of different sizes. RFLPs are very informative due to their codominant nature (they distinguish between homozygotes and heterozygotes). Large numbers of RFLP markers are available for many species. Disadvantages are that they are relatively expensive and time consuming, and they require a large quantity of DNA. Most commonly, radioactive labeled probes are used, but non-radioactive methods are also available.

RH Relative humidity

Rhizobium Genus of bacteria that lives symbiotically with roots of leguminous plants; during the symbiosis, atmospheric nitrogen gas is converted into a form useable by the plant.

Rhizobium nodules Galls on roots caused by Rhizobium spp.

Rhizoid A hairlike unicellular or filamentous absorbing and anchoring structure of organisms in the Kingdom Plantae. Root hairs are a type of rhizoid.

Rhizomatous Having or looking like a rhizome

Rhizome An elongated underground stem, usually horizontal, capable of producing new shoots and roots at the node

Rhizomorph Macroscopic rope-like strand of compacted tissue formed by certain fungi.

Rhizosphere The area around plant roots that is the site of many key interactions between microbes and the plant. It is also the site where nutrients are transformed for take-up by the plant.

Rhizosphere competent Used to describe microorganisms adapted to living in the rhizosphere of a plant.

Rhodopsin A G protein-coupled photoreceptor in retinal rod cells that activates transducin in response to light absorption.

Ribonucleic acid (RNA) Several nucleic acids composed of repeating units of ribose (a sugar), a phosphate group, and a purine (adenine or guanine) or a pyrimidine (uracil or cytosine) base; transcribed from DNA and involved in translation to proteins.

Ribosomal RNA (rRNA) The RNA component of ribosomes.

Ribosome A subcellular protoplasmic particle, made up of one or more RNA molecules and several proteins, involved in protein synthesis.

Ribozyme An RNA enzyme.

Rill A small, intermittent water course with steep sides; usually only several centimeters deep.

Rill erosion The erosion process in which numerous small channels of less than to maybe two to three inches deep are formed.

Ringspot Disease symptom characterized by yellowish or necrotic rings enclosing green tissue, as in some plant diseases caused by viruses.

Riparian zones Thin strips and patches of land and vegetation that border streams.

RNA (ribonucleic acid) A single-stranded nucleic acid similar to DNA but having a uracil rather than thymine as one of the nucleotides. The RNA strand carries the coded information from the nucleus to the cylasm where protein production occurs.

RNA editing RNA processing events other than splicing that alter the protein coding sequences of mRNAs.

RNA polymerase An enzyme that catalyzes the synthesis of RNA by copying the nucleotide sequence of the DNA.

RNA splicing The joining of exons in a precursor RNA molecule.

RNase H An enzyme that degrades the RNA strand of RNA-DNA hybrid molecules.

RNase P A ribozyme that cleaves the 5′ end of pre-tRNAs.

Rogue To remove and destroy individual plants that are diseased, infested by insects, or otherwise undesirable.

Root A vascularized organ of plants that grows into the substrate.

Root (fine) The underground component of plants, responsible for anchoring plants and providing them with water and nutrients. Fine roots, the subject of many research projects, are the very small ephemeral roots (less than 1 mm in diameter) that are largely responsible for providing plants with water and nutrients.

Root cap A group of cells on a root that protects the growing tip.

Root exudates The various compounds that leak from growing and expanding sections of roots as well as from broken cells at exit points of lateral roots.

Root graft The fusion of roots of two adjacent plants so that their water and food conducting (vascular) systems become joined.

Root hair Threadlike, single-celled outgrowths from a root epidermal cell.

Rootstock Portion of the stem (trunk) and associated root system into which a bud or scion is inserted in grafting; fleshy overwintering part of a herbaceous perennial plant with buds and eyes .

Rosette Disease symptom characterized by short, bunchy growth habit due to shortened internodes and no comparable reduction in leaf size.

Rot Softening, discoloration, and often disintegration of plant tissue as a result of fungal or bacterial infection.

Rotation A cropping system in which crops are grown in a field in a fixed annual sequence. A rotation reduces the build up of diseases and pests, aids weed control, improves soil fertility, spreads the risk of crop failure and allows even distribution of labour requirements.

Rotational grazing The practice of of imposing a regular sequence of grazing and rest on the area to be grazed

Rough endoplasmic reticulum (rough ER) A vast system of interconnected, membranous, infolded and convoluted sacks that are located in the cell's cytoplasm (the ER is continuous with the outer nuclear membrane). Rough ER is covered with ribosomes that give it a rough appearance. Rough ER transport materials through the cell and produces proteins in sacks called cisternae (which are sent to the Golgi body, or inserted into the cell membrane).

Roundup A herbicide that provides non-selective control of several annual and perennial weeds. Roundup will also damage crops, such as corn and soybeans that are not Roundup Resistant.

Roundup Ready A trade name given to certain varieties of corn or soybean which are resistant to the herbicide, Roundup, whose active ingredient is glyphosate.

Roundup resistant A plant that is not harmed by the application of Roundup herbicide (i.e. Roundup Ready Corn, Roundup Ready Soybeans).

Rous sarcoma virus (RSV) An acutely transforming retrovirus, in which the first oncogene was identified.

rRNA (abbr. for ribosomal RNA) RNA molecules forming part of the ribosomal structure.

RT-PCR Real time PCR. This is a term used to describe PCR methods which detect DNA amplification in "real time" as the experiment is taking place, rather than at the end of the experiment.

Rugose Wrinkled, roughened.

Ruminant An animal thet chews the cud and has a complex digestive system with a four part stomach enabling bacteria to break down food. Ruminants lack upper incisor teeth and their complex stomach allows them to store and digest large amounts of bulky and fibrous food

Runner plants New plants produced asexually on a runner or stolon.

Runner (stolon) A slender horizontal stem that grows close to the soil surface.

Runoff The portion of precipitation or irrigation on an area that does not infiltrate but, instead, is discharged from the area. That which is lost without entering the soil is called "surface runoff."

Russet Brownish, roughened areas resulting from cork formation.

Rust A disease caused by a specialized group of basidiomycetes that often produces spores of a rusty color.

Ryanodine receptors Calcium channels in muscle and nerve cells that open in response to changes in membrane potential.

S

S phase The phase of the cell cycle during which DNA replication occurs.

Sabadilla A broad-spectrum, short residual (2 days in sunlight) botanical insecticide derived from a South American lily. It can be very toxic to humans in purified form.

Sabulous Sandy; gritty.

Saccate Furnished with or having the form of a bag, pouch, or sac, as a saccate petal.

Sacciform Resembling a pouch.

Safener A chemical agent that reduces toxicity of herbicides to crop plants by a physiological or molecular mechanism; also known as antidotes, antagonists or crop protectants.

Sagittate (of leaves) Derived from sagitta, an arrow; means the leaves are shaped like arrowheads.

Salinity The amount of soluble salts in a soil. The conventional measure of soil salinity is the electrical conductivity of a saturation extract, usually in units of deciSiemens per meter (dS m-1).

Salps Transparent, barrel-shaped planktonic marine animals found singly or in chains that sometimes reach 20 feet in length

Salsuginous Growing naturally in soils with a high salt content; halophytic.

Saltation The movement of soil and mineral particles by intermittent leaps from the ground when particles are being moved by wind or water.

Salverform Noting a corolla with the outer edge spreading out flat, as in the primrose and phlox; hypocrateriform.

Samara A dry, usually one-seeded fruit provided with a wing, as in the ash, elm, or birch. Also called key and key fruit.

Samariform Having the form of a samara.

Sand Individual rock or mineral fragments in a soil that range from 0.05 to 2.0 millimeters in diameter. Most sand grains consist of quartz, but they can be of any mineral composition. Sand is also the textural class name of any soil that contains 85 percent or more sand and no more than 10 percent clay.

Sandy Loam Texture A classification used to describe soil texture. A sandy loam is a soil containing much sand but which has enough silt and/or clay to make it somewhat coherent. The individual sand grains can readily be seen and felt.

Sanitation Destruction or removal of infected and infested plants or plant parts; decontamination of tools, equipment, containers, work space, hands, etc.

Sap transmission Transmission, usually of viruses, by rubbing sap from an infected plant onto a healthy plant to cause infection.

Saponins A group of compounds (triterpene or steroid glycosides) which are produced naturally in a variety of plant species. Saponins are distinctive for their soap-like nature, and they have also been shown to demonstrate various favorable bioactivities.

Saprobe Organism that obtains nourishment from non-living organic matter.

Saprophyte A plant which obtains all its nutrients from decaying matter; this may be an animal or vegetation. Saprophpytic flowering plants usually have no green color in their leaves, which are usually reduced to scales. The roots of these plants are covered by a cobweb-like growth called fungus mycelium. This isn't harmful to the plant, but helps it obtain nutrients from the decaying matter. There are a few flowering plants that are saprophytic; Monotropa and Ericceae are two. There are many saprophytic bacteria and fungi that obtain their nutrients from decaying matter and help the decaying process. The cultivated Mushroom is an example of a saprophyte as it derives its nourishment from decaying manure. A few saprophytic fungi are the causes of disease in plants.

Sapwood Physiologically active zone of wood contiguous to cambium .

SAR (Systemic Acquired Resistance) A long-lasting and broad host-range defense mechanism in plants that can be brought about by previously treating a plant, or parts of it, with microbes that induce cell death.

Sarcocarp The fleshy part of certain fruits, between the epicarp and the endocarp; mesocarp; the part of fleshy fruits usually eaten, as in the peach, plum, etc.

Sarcoma A cancer of cells of connective tissue.

Sarcomere The contractile unit of muscle cells composed of interacting myosin and actin filaments.

Sarcoplasmic reticulum A specialized network of membranes in muscle cells that stores a high concentration of Ca_2+.

Sarment A runner; a slender running stem, as that of the strawberry.

Sarmentose Having long sarments or runners; having the form or character of a runner.

Sartage The clearing of woodland for agricultural purposes, as by setting fire to the trees.

Saturation Soil conditions where all the pores are completely filled with water. The soil tension exhibited on water under this condition is zero, and gravitational is the dominant force moving water downward.

Saxifragous Growing in the crevices of rocks and eventually splitting the rock.

Scab Roughened, crustlike diseased area on the surface of a plant organ.

Scald A necrotic condition in which tissue is usually bleached and has the appearance of having been exposed to high temperatures.

SCAR Sequence characterized amplified region. To improve their specificity and repeatability, a RAPD or AFLP marker can be converted into a SCAR marker. SCARs are developed by obtaining the DNA sequence of a RAPD or AFLP band and designing 18-25 base PCR primers to amplify the same DNA segment. SCARs are easier to employ in marker-assisted selection than RAPDs or AFLPs.

Scion Portion of a shoot used for grafting onto the root stock .

Sclerenchyma cells Plant cells characterized by thick cell walls that provide structural support to the plant.

Sclerotium A hard, resistant vegetative resting body of a fungus composed of a compact mass of hyphae and capable of surviving under unfavourable environmental conditions. Under favourable conditions the sclerotium can produces sexual or asexual fruiting bodies.

Scorch Any symptom that suggests the action of flame or fire on the affected part, often seen at the margins of leaves.

Scrapie Long established disease of sheep that causes intense irritation: cause unknown, thought to be neither bacteria or virus.

Scutellar epithelium A continuous layer of cylindrical cells covering the endosperm side of the scutellum Thought to act as a secretory and absorptive organ.

Scutellum A shield-like structure which represents the monocotyledon of the grass embryo It is pressed against the endosperm and functions as an absorptive organ.

SDS-polyacrylamide gel electrophoresis (SDS-PAGE) A commonly used method to separate proteins by gel electrophoresis on the basis of size.

Seafloor speading Area where two tectonic plates are moving apart (diverging), opening the seafloor and allowing magma to rise and form new oceanic crust.

Seamount An extinct, underwater volcano that rises more than 1 km above the seafloor but whose peak is below the sea surface.

Second messenger A compound whose metabolism is modified as a result of a ligand-receptor interaction; it functions as a signal transducer by regulating other intracellular processes.

Second polar event The fusion of a male gamete with two haploid female nuclei at the time of the fertilisation of the ovule The resulting triploid nucleus divides to form the endosperm

Secondary embryogenesis The development of embryos from young embryos.

Secondary infection Infection resulting from the spread of infectious material produced after a primary infection or from secondary infections without an intervening inactive period.

Secondary Inoculum inoculum produced by infections that took place during the same growing season. Compare with primary inoculum.

Secondary messengers Transient secondary signals inside the cell that greatly amplify the original signal.

Secondary Metabolite A compound that is not necessary for growth or maintenance of cellular functions but is, in general, synthesised for the protection of a cell or microorganism during the stationary phase of its growth cycle. In plants, secondary metabolites are believed to be important in the attraction of pollinators and to have a function in defense, as some metabolites have antimicrobial properties. Also known as secondary products.

Secondary organism Organism that multiplies in already diseased tissue but is not the primary pathogen.

Secondary pollutant Air pollutant that must be chemically produced from other air pollutants, e.g. ozone (O3) that is a product of a photochemical reaction of exhaust products from combustion engines in the atmosphere.

Secondary root Branch from a primary root.

Secondary structure The regular arrangement of amino acids within localized regions of a polypeptide chain. See a helix and b sheet.

Secretory vesicles Membrane-enclosed sacs that transport proteins from the Golgi apparatus to the cell surface.

Sedentary Remaining in a fixed location.

Sediment Eroded particles that are no longer part of the soil. These sediments, sand silt, or clay size, are carried by wind or water and deposited to a different area, like surface water bodies. Sedimentation occurs when eroded soil is deposited and/or settles in water. Erosion, itself, is the process of detaching and transporting soil particles.

Sedimentary rock Rock formed from the weathered products of preexisting rocks that have been transported by water, wind, or glaciers. Sandstone is a type of sedimentary rock.

Sedimentation Occurs when eroded soil deposits and/or settles out in water.

Seed A multicellular structure containing the embryo of a seed plant, ordinarily with stored food, the whole protected by a seed coat.

Seed treatment Application of a biological agent, chemical substance, or physical treatment to seed, to protect the seed or plant from pathogens or to stimulate germination or plant growth.

Seedborne Carried on or in a seed.

Seedling growth inhibitors Mode of action of a group of herbicides that generally disrupt early growth of plants.

Seedling growth inhibitors Mode of action of a group of herbicides that generally disrupt early growth of plants.

Seep A small area where water ñ that may be of a different temperature and density flows from below the seafloor and rises slowly into the ocean.

Segregating population A colelction of lines from a cross or population that is undergoing Mendelian or Hardy-Weinberg gene segregation.

Segregation distortion Deviation of the expected, segregation ratio from that expected by Mendelian inheritance. Distorted segregation means that either the maternal or paternal allele was transmitted to the progeny at a higher frequency than expected.

Seismic wave Waves generated either by an earthquake or artificially.

Selectable marker gene A gene encoding an easily detectable trait transferred into a cell with another gene encoding a less detectable trait. The selectable marker gene allows genetic engineers to select transgenic cells and plants from those that are non-transgenic.

Selectins Cell adhesion molecules that recognize oligosaccharides exposed on the cell surface.

Selection media Agarose media containing the substance for which the selectable marker gene encodes resistance. When tissue culture cells are place on this media, cells that have not been transformed and do not contain the gene conferring resistance to the substance will die leaving only those cells that are transgenic.

Selection pressure The process that changes the relative frequency of one or more genes within a population.

Selective Herbicides which cause one species of plants to die, but another species will survive the same treatment. Selectivity allows for acceptable weed control without injury to crop plants.

Selective grazing Taking certain plant species, individual plants, or plant parts to the exclusion of others.

Selective medium A culture medium containing substances that specifically inhibit or prevent the growth of some species of microorganisms.

Selectividad Lmplica que diferentes especies vegetales no responden de la misma manera a un herbicida particular. Los herbicidas que causan la muerte de una especie de plantas pero no la muerte de otra especie son herbicidas selectivos. La selectividad permite un control aceptable de malezas sin dañar a los cultivos.

Selectivity Differential effect of a herbicide when applied to a mixed population of plants; sensitive plants die, tolerant plants are unharmed. Implies that different plant species do not respond in the same way to a particular herbicide. Herbicides which cause one species of plants to die, but not another, are selective. Selectivity allows for acceptable weed control without injury to crop plants.

Self-fertilization A process in which sexual reproduction occurs as a result of the fusion of sex cells produced by the same individual.

Self-splicing The ability of some RNAs to catalyze the removal of their own introns.

Seminal root A root of a vascular plant that arises from the radicle or the zone of lateral root formation.

Senesces The end of a plant's life cycle when it reaches maturity and leaves dry up.

Sepals A modified leaf which is usually green and part of the caylx; can protect the flower when in bud.

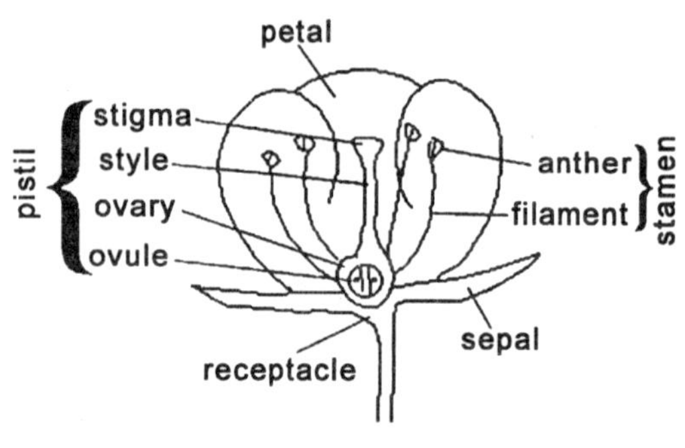

Fig. Sepal

Septate With cross walls; having septa.

Septum Dividing wall; in fungi, cross wall.

Sequencing Determination of the nucleotide sequence of the DNA. DNA is composed of 4 nucleotides represented by the 'building blocks' A, C, G, and T. The order of A, C, G and T within the DNA strand determines the start and s, when the gene is expressed and most importantly provides the information specifying the amino acid sequence of the protein to be produced from that gene.

Serology A method using the specificity of the antigen-antibody reaction for the detection and identification of antigenic substances and the organisms that carry them.

Serrate Edges with teeth, like a saw.

Sessile Used in reference to a leaf, leaflet, flower, floret, fruit, ascocarp, basidiocarp, etc., without a stalk, petiole, pedicel, stipe or stem; (of nematodes) permanently attached; not capable of moving about.

Set aside Set aside was introduced as part of a programme for tackling the over production of cereals within the EU. A major review of the CAP in 1992 (MacSharry reforms) addressed over production and agreed to reduce the amount of subsidies paid to farmers for cereals by 35% over the proceeding three years. To help compensate farmers for their loss of income the Arable Area Payments Scheme was introduced. Under this scheme, farmers can claim support payments based on the size of the area used to grow cereals, linseeds, oilseeds and protein crops such as peas, beans and lupins. However, to qualify for these payments, all but the smallest producers must 'set aside' part of their arable land, taking it out of production.

Set stocking Grazing systems in which stock remain in one field or paddock for a long period.

Seta (setae) The stalk that bears the capsule of the sporophyte of a moss or liverwort. Any bristle-like structure.

Severity values A means to quantify accumulating opportunities for pathogen infection (e.g. infection periods) to a pre-determined threshold that requires a disease management activity (e.g. a fungicide application); used in disease prediction or forecasting.

Sexual phase The portion of an organism's life cycle devoted to sexual reproduction.

Sexual reproduction Reproduction involving fusion of two haploid nuclei (karyogamy) to form a diploid nucleus followed by meiosis (reduction division) back to haploid nuclei at some point in the life cycle, resulting in genetic recombination.

Sexual spore Spore produced during the sexual cycle.

Sexually compatible Able to be cross-mated or cross-fertile.

SH2 domain A protein domain of approximately 100 amino acids that binds phosphotyrosine-containing peptides.

Shattercane Sorghum bicolor, is a summer annual that resembles grain sorghum and corn. It is derived from wild and/or cultivated sorghum varieties. Shattercane is a major weed problem in corn, sorghum, and soybean fields in the Central United States.

Shear Strength The ability of a soil or rock to resist shearing (tangential) forces and/or stresses.

Sheath The flattened base of a leaf that wraps around a stem.

Sheet erosion The removal of a relatively uniform thin layer of soil for the land surface by unchanneled runoff, or sheet flow

Shikimate pathway Important chemical pathway in plants. It is responsible for producing amino acids and other aromatic products.

Shine-Delgarno sequence The sequence prior to the initiation site that correctly aligns bacterial mRNAs on ribosomes.

Shoot A portion of a plant consisting of a stem and its attached leaves.

Shoot apex See apical meristem

Shot hole Symptom in which small lesions fall out of leaves, giving the leaf the appearance of being hit by buckshot.

Shrink-Swell Potential The susceptibility of a soil to a change in volume due to a loss or gain in moisture, especially those soils with a relatively high clay content.

Sieve element (sieve tube element) A tube-shaped living cell in the phloem functioning in the transport of dissolved organic substances in the plant.

Sign Indication of disease from direct observation of a pathogen or its parts .

Signal patch A recognition determinant formed by the three-dimensional folding of a polypeptide chain.

Signal peptidase An enzyme that removes the signal sequence of a polypeptide chain by proteolysis.

Signal recognition particle (SRP) A particle composed of proteins and 7SL RNA that binds to signal sequences and targets polypeptide chains to the endoplasmic reticulum.

Signal sequence A hydrophobic sequence at the amino terminus of a polypeptide chain that targets it for secretion in bacteria or incorporation into the endoplasmic reticulum in eukaryotic cells.

Signal-transduction pathway The process whereby a receptor initiates one or more sequences of biochemical reactions that connect the stimulus to a cellular response.

Silt A soil inorganic separate in the range of 2 to 50 micrometers (or 0.002 to 0.05 mm.). Silt is smaller than sand but larger than clay.

Silviculture The care and cultivation of forest trees; forestry.

Single gene resistance (monogenic resistance) Resistance conferred by a single gene.

Single seed descent Breeding procedure in which progenies of a segregating population are self-pollinated and advanced to the next generation by a single seed in each progeny line.

Singlet states An orbital to which an electron may move after being excited by a photon.

Singlet state orbitals require that the excited electron have a spin opposite to that of the spin on the electron remaining in the ground state.

Siphonostele A stele consisting of a cylinder of vascular tissue surrounding a pith; the cylinder may be interrupted by parenchyma-filled leaf traces.

Sister taxa Monophyletic taxa that are each others closest relatives, i.e., they share an immediate common ancestral species.

Site of Absorption Site of Absorption is the location where the herbicide is taken up by the plant. This should not be confused with Site of Action which is the biochemical pathway within the plant where the herbicide acts.

Site of Action The Site of Action is the biochemical pathway a particular herbicide acts upon in a plant. Currently there are over 20 sites of action

Site of Uptake Site of Uptake is the location where the herbicide is taken up by the plant. This should not be confused with Site of Action which is the biochemical pathway within the plant where the herbicide acts.

Soil development Soil development is influenced by five factors: parent material, climate, living organisms, topography, and time.

Soil classification A specific soil is classified according to the number of horizons in its soil profile and the soil properties of each horizon. Well developed soils are old and contain all three master horizons (A, B, and C), as well as several subdivisions of the master horizons.

Soil air The soil atmosphere; the gaseous phase of the soil, being that volume not occupied by solid or liquid.

Soil aggregation/aggregates The process of soil particles binding together. The mineral components of soil are clay, silt and sand. Clay particles are too small to be seen with a light microscope, and sand particles can be seen with the naked eye. Silt particles are of intermediate size. Without soil aggregation, soils would not have pores large enough for movement of air and water, and roots would not be able to penetrate the soil.

Soil age Determined by the amount of weathering that has occurred; to what extent the parent material has been converted to distinct horizons or soil layers. Usually described as young, mature or old.

Soft rot Softening, discoloration, and often disintegration of plant tissue as a result of fungal or bacterial infection.

Soft dough A description the grain at maximum fresh weight. When punctured the grain contents are still wet but starting to become floury.

SNP Single Nucleotide Polymorphism. A single position in a given DNA sequence that contains allelic variations within a population at relatively high frequencies. For example, replacement of a guanine (G) with a thymine (T) nucleotide is an example of a SNP. SNPs occur often in many genomes, thus facilitating the development of high-density marker maps.

Smut A disease caused by a smut fungus (Ustilaginales) in the Basidiomycota or the fungus itself; it is characterized by masses of dark brown or black, dusty to greasy masses of teliospores that generally accumulate in black, powdery sori.

Smooth endoplasmic reticulum (smooth ER) A vast system of interconnected, membranous, infolded and convoluted tubes that are located in the cell's cytoplasm (the ER is continuous with the outer nuclear membrane). The space within the ER is called the ER lumen. Smooth ER transport materials through the cell. It contains enzymes and produces and digests lipids (fats) and membrane proteins; smooth ER buds off from rough ER, moving the newly-made proteins and lipids to the Golgi body and membranes

Smectite A group of silicate (i.e., rocks or minerals with their primary unit as silicon tetroxide, SiO4) clay minerals that have shrink-swell property.

Slurry A semi fluid mixture of faeces and urine which may also include rain water and yard washings.

Slime molds (Myxomycetes) Saprophytic organisms that form vegetative amoeboid plasmodia and spores.

Site specific recombination Recombination mediated by proteins that recognize specific DNA sequences.

Soil The layer of the Earth's surface, consisting of four major components: air, water, organic matter, and mineral matter . There are three categories of soil particles—sand, silt, and clay—which are called "soil separates."

Soil drench Application of a solution or suspension of a chemical to the soil, especially pesticides to control soilborne pathogens.

Soil forming factors The variables, usually interrelated natural agencies, that are active in and responsible for the formation of soil. The factors are usually grouped into five major categories as follows: parent material, climate, organisms, ography, and time.

Soil horizons Soils are generally composed of many different layers, or horizons lying on top of one another. A variety of biological (e.g., decomposition), chemical (e.g. rain/soil interactions), and physical (e.g., weathering) processes form these horizons. The O (organic) horizon is closest to the surface, with mineral horizons such as A, E, B, and C lying underneath the O horizon and above bedrock.

Soil inhabitant An organism that maintains its population in soil over a period of time.

Soil invader An organism whose population in soil diminishes in several months to years.

Soil organic matter The organic fraction of the soil, exclusive of undecayed plant and animal residues.

Soil pasteurization Process used to free soil of selected harmful microorganisms using heat.

Soil sample A collection of individual cores from a known area.

Soil series A unit of soil classification determined by studying horizon characteristics, such as: number of horizons, color, thickness, texture, erosion phase, slope, organic content, and depth to hardpan. All soils given the same soil series name possess the same characteristics across the landscape.

Soil sterilization Process used to free soil of all microorganisms.

Soil structure The arrangement of soil separates into units called soil aggregates.

Soil taxonomy The classification and naming of soils.

Soil test correlation The relationship between the amount of nutrient extracted from soil by a laboratory test and nutrient uptake by plants in the greenhouse or field and/or crop yield.

Soil test Soil samples are chemically analyzed to assess soil nutrient levels and determine how fertilizer use can be improved.

Soil texture This refers to the relative proportions of sand, silt and clay in a soil.

Soil water content The water lost from the soil upon drying to a constant mass at 105 degrees C., expressed either as the mass of water per unit mass of dry soil or as the volume of water per unit bulk volume of soil.

Soil water deficit (SWD) The amount of water required to restore field capacity

Soilborne Carried on or beneath the soil surface.

Solarization Disease control practice in which soil is covered with polyethylene sheeting and exposed to sunlight, thereby heating the soil and controlling soilborne plant pathogens.

Solubility The ability of a solute to dissolve in a solvent; the ability of a herbicide to dissolve in water.

Soluble A state change when a solid material is put into a solvent. The molecules leave the solid phase and go into the liquid phase of the solvent

Solute molecules Molecules that are dissolved in a solvent

Solute potential The effect of dissolved solutes on water potential

Solution A liquid matrix where a solute has been dissolved in a solvent.

Solvent A liquid that is capable of dissolving other molecules called solutes. In biological systems, water is usually the solvent

Somatic cells Body cells.

Sonar Acronym for sound navigation and ranging. A device that is used primarily for the detection and location of underwater objects by reflecting acoustic (sound) waves from them.

Sooty mold Black, nonparasitic, superficial fungal growth on honeydew produced by aphids and other phloem-feeding insects.

Sorption The removal of an ion or molecule from solution. Commonly used in regards to phosphorus fixation.

Sorus Compact fruiting compact fruiting structure, especially the erumpent spore mass in the rust fungi (Uredinales) and smut fungi (Ustilaginales); occasionally a group of fruiting bodies as in Synchytriaceae; a cluster of sporangia on a fern sporophyte.

Southern Blot A method used to transfer DNA fragments separated during electrophoresis to membrane filters for detection of specific nucleotide sequences using probes.

sp. A genus name followed by sp. means that the particular species is undetermined; spp. after a genus name means that several species are being referred to.

Spatial Of, relating to, involving, or having the nature of space.

Species Any one kind of life subordinate to a genus but above a race; a group of closely related individuals of the same ancestry, resembling one another in certain inherited characteristics of structure and behavior and relative stability in nature; the individuals of a species ordinarily interbreed freely and maintain themselves and their characteristics in nature.

Species diversity The number of different species in a given area.

Specific epithet The second word in a Latin binomial.

Specific resistance (vertical resistance) Resistance which is effective against some biotypes or races of the pathogen, but not others, usually inherited monogenically and expressed qualitatively. (see general resistance, horizontal resistance, race-nonspecific resistance).

Spectrin A major actin-binding protein of the cell cortex.

Sperm The male gamete.

Spermagonium Structure in which male reproductive cells are produced; in rust fungi, globose or flask-shaped haploid fruiting body composed of receptive hyphae and spermatia (pycniospores).

Spermatangium (spermatangia) In gametophytes of the Rhodophyta, the structures that form spermatia; equivalent to antheridium.

Spermatium A male sex cell; a nonmotile male gamete; a haploid male gamete.

Spermatocyte The diploid cell that undergoes meiosis to form sperm.

Spermatogenous tissue groups of (usually haploid) cells that turn into sperm cells.

Sphingomyelin A phospholipid consisting of two hydrocarbon chains bound to a polar head group containing serine.

Spike An unbranched, indeterminate iflorescence with sessile flowers.

Spikelet Spike-like appendage comprised of one or more reduced flowers and associated bracts; unit of inflorescence in grasses; a small spike.

Spiroplasma Spiral-shaped plant pathogenic mollicute (prokaryote without cell wall).

Spliceosomes Large complexes of snRNAs and proteins that catalyze the splicing of pre-mRNAs.

Spodosols Spodosols are soils characterized by an accumulation of aluminum, iron, and organic matter. These soils are often present in cool, moist environments, and represent the soils.

Spongy parenchyma The tissue typically found between the palisade parenchyma and the lower epidermis of leaves. Spongy parenchyma consists of loosely arranged, irregularly shaped cells containing chloroplasts and is interspersed with intercellular spaces.

Spontaneous generation, theory of The theory, now known to be invalid, that plants, animals and microorganisms arose suddenly from non-living materials under certain environmental conditions.

Sporangiophore Sporangium-bearing body of a fungus.

Sporangiospore Non-motile, asexual spore that is borne in a sporangium.

Sporangium Saclike fungal structure in which the entire contents are converted into an indefinite number of asexual spores.

Spore Reproductive structure of fungi and some other organisms, containing one or more cells; a bacterial cell modified to survive an adverse environment.

Sporidium Basidiospore of rust fungi, smut fungi, and other Basidiomycota.

Sporocarp Spore-bearing fruiting body.

Sporocyte The diploid cell that undergoes meiosis to form meiospores.

Sporodochium Superficial, cushion-shaped asexual fruiting body consisting of a cluster of conidiophores.

Sporogenous Spore-producing.

Sporophore A spore-producing or spore-bearing structure such as a conidiophore, ascocarp or basidiocarp.

Sporophyll A specialized leaf or leaf-like structure that bears meiosporangia.

Sporophyte A diploid organism arising from a zygote; produces meiospores by meiosis.

Sporopollenin The inert polymerized carotenoid that forms the outer cell wall of meiospores in the Kingdom Plantae.

Sporulate To produce spores.

Spot A symptom of disease characterized by a limited necrotic area, as on leaves, flowers, and stems.

Springwood Early part of the yearly xylem growth ring in woody plants consisting typically of cells that are larger than those formed later in the season (summerwood).

Spur shoot A lateral shoot with short or nonexistent internodes, sometimes bearing reproductive structures or fascicles of leaves.

Src A nonreceptor protein-tyrosine kinase encoded by the oncogene (src) of Rous sarcoma virus.

SSR Simple Sequence Repeat, also known as a microsatellite. These DNA markers are based on differences in length of repeated di- and tri-nucleotide sequences. By identifying the conserved flanking sequences of these SSRs, primers can be designed to selectively amplify them via PCR. SSRs are usually codominant and a potentially large number can be identified. Only small amounts of DNA are required, and analyses can be run quickly and relatively inexpensively. However, initial identification of SSRs followed by design, synthesis, and evaluation of primers is a large-scale effort.

SSSI Site of Specific Scientific Interest.

Stabilizing selection The theorized competitive disadvantage of unnecessary virulence genes; races with excess genes would have decreased fitness relative to races with fewer virulence genes, so a "super-race" would be less likely to appear in multiline crops.

Staghead Defoliated, dead or dying major branches in the crown of a tree, usually resulting from inadequate water uptake or translocation.

Stalk rot One type of stalk rot (Anthracnose stalk rot) is caused by a fungus (Colletotrichum graminicola) that enters the corn plant above ground. The disease causes the inside of the stalk to degrade from bright white to gray. This disease weakens the stalk making it more likely for plants to lodge.

Stamen The structure in a flower that produces pollen grains, consisting of a stalk (filament) and a terminal lobed head (anther).

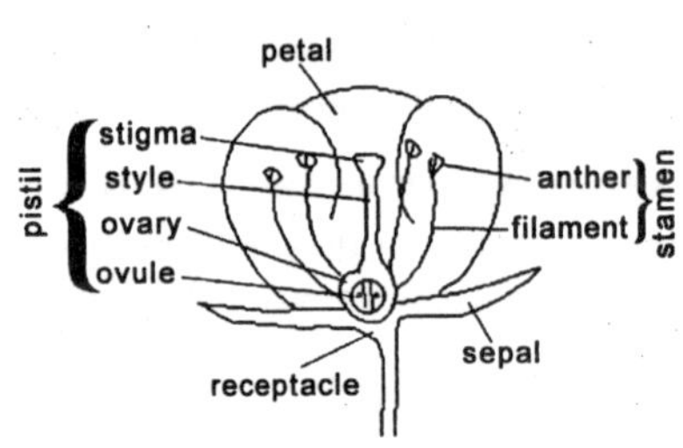

Fig. Stamen

Staminode A stamen without anthers.

Starch A polymer of glucose residues that is the principal storage form of carbohydrates in plants.

START A regulatory point in the yeast cell cycle that occurs late in G1. After this point a cell is committed to entering S and undergoing one cell division cycle.

Start/stop codons Codons used to start and stop translation during protein synthesis.

STAT proteins Trancription factors that have an SH2 domain and are activated by tyrosine phosphorylation, which promotes their translocation from the cytoplasm to the nucleus.

Stele Central cylinder of vascular tissue (especially in roots).

Stem The vascularized above-ground supportive portion of a plant, and below-ground structures with the same anatomy and development.

Stem cell A cell that divides to produce daughter cells that can either differentiate or remain as stem cells.

Stem pitting A viral disease symptom characterized by depressions on the stem.

Stereochemically The specific 3-dimensional arrangement of atoms around an asymmetric carbon.

Stereocilium A specialized microvillus of auditory hair cells.

Sterigma Small, usually pointed projection that supports a spore.

Sterile Unable to reproduce sexually;

Sterile fungus A fungus that is not known to produce any kind of spores.

Sterility Unable to produce seed, pollen or spores capable of fertilization

Sterilization The total destruction of living organisms by various means, including heat, chemicals or irradiation.

Steroid hormones A group of hydrophobic hormones that are derivatives of cholesterol.

Steroid receptor superfamily A family of transcription factors that regulate gene expression in response to steroids and related hormones.

Stigma The glandular sticky surface at the tip of a carpel of a flower, which receives the pollen.

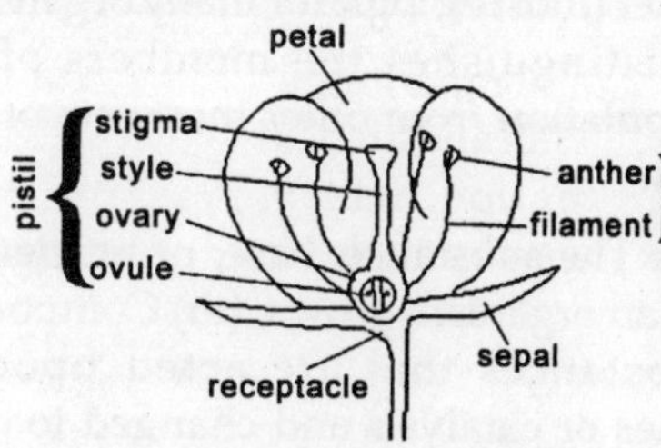

Fig. Stigma

Stipe Stalk.

Stippling Series of small dots or speckles in which chlorophyll is absent.

Stipule Small leaflike appendage at the base of a leaf petiole, usually occurring in pairs.

Stochastic A random variable, or referring to patterns resulting from random effects.

Stocking density The number of animals per unit area of land at a point in time

Stocking rate The number of animals per unit area over a given period

Stolon (runner) A slender, horizontal stem that grows close to the soil surface; in fungi, a hypha that grows horizontally along the surface.

Stoma A minute opening in the epidermis of leaves and stems through which gasses pass. The stoma is bordered by two guard cells which open and close depending on their turgidity. The term stoma is often used to refer to the entire stomatal apparatus i.e. the guard cells as well as the pore they surround.

Stomatal Closure Response The response by plants to environmental stress such as drought, which reduces stomatal pore size. This leads to reduced water transpiration rates and restricted access of CO_2 to the leaves.

Stone fruit Fruit with a stony endocarp, e.g. cherry, peach, plum.

Store cattle (or sheep) Animals which have been grown slowly so that their skeletal development is not impaired, but muscle tissue is slightly below the animals potential and fatty tissue is not developed.

Strain A distinct form of an organism or virus within a species, differing from other forms of the species biologically, physically, or chemically.

Stratification 1. exposing seeds to a cold period to encourange germination or break seed dormancy;

2. layering of plant material (herbacous, shrub, understory, canopy)

Stratified Exposed to a cold period to break seed dormancy

Stream gauge A device used to measure streamflow volume over time.

Striate (striations) Marked with delicate lines, grooves, or ridges.

Strobilus (strobili) A cone; a compact group of meiosporangium-bearing structures.

Stroma Compact mass of mycelium (with or without host tissue) that supports fruiting bodies or in which fruiting bodies are embedded.

Structure The combination or arrangement of soil particles that forms peds or aggregates.

STS Sequence tagged site. A DNA sequence of known location that occurs just once in the genome. STS loci are detectable by PCR, making them useful for a variety of molecular marker applications.

Stubble The part of a crop left after harvesting

Stunting Reduction in height of a vertical axis resulting from a progressive reduction in the length of successive internodes or a decrease in their number.

Style Part of the pistil (female reproductive structure) that transports the sperm collected by the stigma to the ovary for fertilization and developement of the egg; situated between the ovary and stigma.

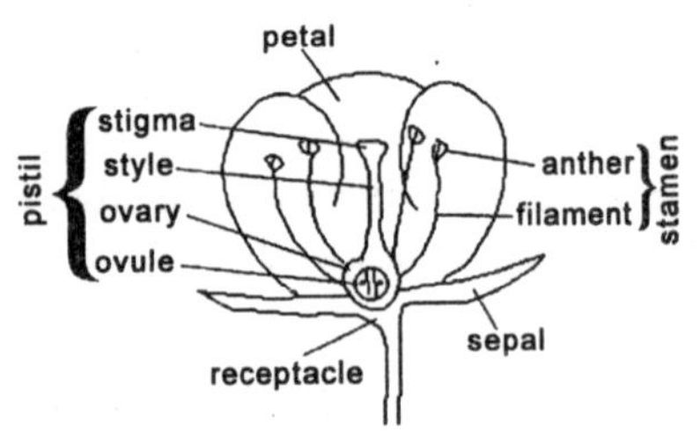

Fig. Style

Stylet Stiff, slender, hollow feeding organ of plant-parasitic nematodes or sap-sucking insects, such as aphids or leafhoppers.

Stylet knob (basal knob) Structure at the base of a nematode stylet.

Stylet-borne transmission (nonpersistent transmission) A type of virus transmission in which the virus is acquired and transmitted by the vector after short feeding times, and is retained by the vector for only a short period of time.

Subduction zone The region where one lithospheric plate descends beneath another as the two plates are moving towards each other. It is characterized by a line of earthquakes that demarcate the upper edge of the descending plate.

Suberin A waterproofing material secreted by cells of endodermis or periderm.

Suberize To convert into cork tissue.

Subgenomic RNA A piece of viral RNA, shorter than the entire genome of the virus, found in cells infected by the virus and sometimes encapsidated.

Subsample A small portion of an homogenous sample. For example, scientists interested in historical qualities of precipitation may take a 100 mL subsample of an archived 1 L precipitation sample.

Subspecies A subpopulation of a species, defined on the basis of more than one character (morphologic for many organisms) that distinguishes the members of the subpopulation from other members of that species.

Substrate The substance, base, or nutrient on which an organism grows, (iii) Compounds or substances that are acted upon by enzymes or catalysts and changed to other compounds in the chemical reaction.

Substrate analogs Molecules that differ in structure, at one or more positions, from the endogenous substrate.

Succession Process in which communities of plant and animal species in a particular area are replaced over time by a series of different and usually more complex communities. There are two types of succession: primary and secondary. Primary succession occurs in a bare area that has never been occupied by a community of organisms (for example, on a newly-cooled lava flow, or newly-risen oceanic island), and secondary succession occurs in an area in which natural vegetation has been removed or destroyed but the soil has not been destroyed (for example, after forest logging).

Suicide substrate A substrate which irreversibly binds to the active site of an enzyme, causing it to be unable to catalyze further reactions.

Sulfonylureas A herbicide family with similar chemical structure belonging to the amino acid synthesis inhibitors mode of action; these herbicides specifically target the ALS-AHAS enzyme.

Sulfur dioxide (SO2) A primary air pollutant produced in industrial processes and coal burning that causes interveinal necrosis on broadleaf plants and tip necrosis on conifers.

Summerwood Part of the yearly xylem growth ring in woody plants formed late in the growing season and consisting of cells smaller than those of springwood.

Sunscald or sunburn Injury of plant tissues burned or scorched by direct sun or or sunburn.

Superphosphoric acid Made by dehydrating water from phosphoric acid; a mixture of polyphosphate and orthophosphate; used to produce ammonium polyphosphate fertilizers.

Suppressive soil Soil in which various diseases are naturally at lower levels than expected due to biological factors in the soil; an example of natural biological control.

Surface Creep The creeping of the surface.

Surfactant An additive that reduces the surface tension ofa herbicide to allow for better contact between water droplet and plant surface; cationic surfactants, ionic surfactants, nonionic surfactants.

Suscept An abbreviated term for a susceptible plant.

Susceptibility Is the lack of an ability to withstand the herbicide at normal use rates and the plant dies.

Susceptible Prone to develop disease when infected by a pathogen (see resistance).

Suspension The containment or support in fluid media (usually air or water) of soil particles or aggregates, allowing their transport in the fluid when it is flowing. In fluids at rest, suspension follows Stoke's Law. In wind this usually refers to particles or aggregates <0.1 mm diameter through the air, usually at a height of >15 cm above the soil surface, for relatively long distances.

Suspensor The organ anchoring and supporting a developing embryo. Suspensor connects embryo with its mother body.

Sustainability from Sustain meaning: To hold up; to bear; to support; to provide for; to maintain; to sanction; to keep going; to keep up; to prolong; to support the life of. (Chambers Concise Dictionary)

Sustainable agriculture The application of husbandry experience and scientific knowledge of natural processes to create integrated, resource conserving farming systems, based on respect for the people and animals involved which reduce environmental degradation, and which promote agricultural productivity and economic viability in both the short and long term.

Symbiosis A close, prolonged association between two or more different organisms of different species that may, but does not necessarily, benefit each member. The relationship between nitrogen-fixing bacteria and roots of legumes is an example of a symbiotic relationship. The bacteria live in the roots of these legumes, and convert atmospheric nitrogen gas (not useful to plants) to nitrate, a plant nutrient. The bacteria benefit by having a good habitat in which to live, and the plants benefit by using the produced nitrate. Many other types of roots have symbiotic relationships with certain kinds of fungi.

Symbiotic algae Algae (a simple type of plant that can photosynthesize) that live together with another organism in a mutually beneficial relationship.

Symplastic path Water travels from one cell to the next via plasmodesmata. Symplastic means 'living' tissue.

Sympodial Pertaining to proliferation of axes, in which each successive spore or branch develops behind and to one side of the previous apex where growth has ceased.

Symport The transport of two molecules in the same direction across a membrane.

Symptom Indication of disease by reaction of the host, e.g. canker, leaf spot, wilt

Symptomless carrier A plant that, although infected with a pathogen (usually a virus), produces no obvious symptoms.

Synangium (synangia) a structure formed of two or more cohering sporangia.

Synapse The junction between a neuron and another cell, across which information is carried by neurotransmitters.

Synaptic vesicle A secretory vesicle that releases neurotransmitters at a synapse.

Synaptonemal complex A zipperlike protein structure that forms along the length of paired homologous chromosomes during meiosis.

Syncytial endosperm A mass of cytoplasm with several nuclei, not divided into separate cells.

Syncytium A multinucleate structure in root tissue formed by dissolution of common cell walls induced by secretions of certain sedentary plant-parasitic nematodes, e.g. cyst nematodes.

Synergism Increased phytotoxicity occurs when two or more chemicals are applied together as compared to the individual chemicals applied alone.

Synergist A non-phytotoxic chemical which increases the phytotoxicity of a herbicide, usually by inhibiting metabolism of the herbicide through enzyme inhibition.

Syngamy The union of gametes to form the zygote; fertilization.

Synnema Compact or fused, generally upright conidiophores, with branches and spores forming a headlike cluster.

System A number of components linked together

Systematics The study of the kinds of organisms and the relationships between them.

Systemic 1. Pertaining to a disease in which the pathogen spreads generally throughout the plant ;

2. pertaining to a chemical absorbed into the plant through root or foliage and transported internally throughout the plant.

Systemic acquired resistance (SAR) Reduced disease symptoms on a portion of a plant distant from the area where a hypersensitive response occurred or other stimulus was applied; a rapid and coordinated defense response against a variety of pathogens as a signal travels throughout the plant .

Systemic fungicide A chemical agent that spreads internally through the plant and eradicates established fungal infections. See fungicide.

Systemic fungicide A fungicide that is absorbed into plant tissue and may offer some curative or after-infection activity; includes fungicides that are locally systemic, xylem-mobile (upward moving), and amphimobile (move in phloem upward as well as downward in the plant) .

Systemic Having the ability to move throughout the plant by either the xylem or phloem.

Systemic Having the ability to move throughout the plant by either the xylem or phloem.

Systemic herbicide A herbicide that can move throughout the plant; the activity is not limited to the point of application.

T

T cell receptor A T lymphocyte surface protein that recognizes antigens expressed on the surface of other cells.

Tapetum (tapeta) A special tissue surrounding the microsporocytes in the anthers of Anthophyta.

Taproot Primary root that grows vertically downward and from which smaller lateral roots branch.

Taq polymerase A DNA polymerase enzyme which is highly heat stable, used in PCR procedures.

Taqman A real time PCR probe detection system, using two labelled probes.

Target DNA The particular portion of the template DNA which is copied during PCR.

TATA box A regulatory DNA sequence found in the promoters of many eukaryotic genes transcribed by RNA polymerase II.

TATA-binding protein (TBP) A basal transcription factor that binds directly to the TATA box.

Taxol A drug that binds to and stabilizes microtubules.

Taxon (taxa) a group of organisms in a taxonomic hierarchy that have affinity to one another.

Taxonomy The science dealing with naming and classifying organisms.

Teleomorph (perfect state) The sexual form in the life cycle of a fungus .

Teliospore (sometimes called teleutospore, teleutosporodesm) Thick-walled resting or overwintering spore produced by the rust fungi (Uredinales) and smut fungi (Ustilaginales) in which karyogamy occurs; it germinates to form a promycelium (basidium) in which meiosis occurs .

Telium Fruiting body (sorus) of a rust fungus that produces teliospores.

Telomerase A reverse transcriptase that synthesizes telomeric repeat sequences at the ends of chromosomes from its own RNA template.

Telomeres Repeats of simple-sequence DNA that maintain the ends of linear chromosomes.

Telophase The final phase of mitosis, during which the nuclei re-form and chromosomes decondense.

Temperate Characterized by moderate temperatures, weather, or climate; neither hot nor cold.

Temperature sensitive mutant A cell expressing a protein that is functional at one temperature but not at another, whereas the normal protein is functional at both temperatures.

Temporal Of, relating to, or limited by time. Research is often conducted to look at "temporal patterns," or those patterns that occur and change over time. A "temporal scale" refers to a scale of time (e.g., a scale of many years or months).

Temporary wilt Wilt due to insufficient soil water from which a plant can recover when water is supplied.

Tepal An element of a perianth that is not clearly either a sepal or a petal.

Teratogen A chemical that causes malformations in the fetus.

Terminal An accumulation of soil, pebbles, rocks, and boulders carried forward, deposited and marking the farthest advance of a glacier or ice sheet.

Terminal spikelet This marks the completion of the spikelet initiation phase in wheat. The terminal spikelet has its plane of symmetry at right angles to the other spikelets.

Terminal tip On the end of a branch or stem

Termination sequence The sequence of DNA which signals the transcription to stop.

Terminator genes A set of transgenes designed to arrest the final stages of seed development in a crop plant. The resulting trait is the production of seeds that will not germinate. Also called the technology protection system.

Terraces Physical earthen barriers to runoff and can be very effective in reducing P losses to erosion and runoff.

Tertiary structure The three-dimensional folding of a polypeptide chain that gives the protein its functional form.

Test A hard shell produced by some unicellular protists; may be made of calcium carbonate, silica, or sand grains.

Testa The outer coat of the seed, formed by the hardened integuments

Testosterone A steroid hormone produced by the testis.

Tetrad A group of four meiospores.

Tetragonal tetrad A tetrad in which the meiospores are arranged in a square.

Tetrahedral tetrad A tetrad in which the meiospores are arranged in a tetrahedron.

Tetrapyrrole A class of molecules characterized by a series of four, five-membered rings containing four carbon atoms and one nitrogen. Tetrapyrrole molecules include hemes, chlorophylls, vitamin B12 and phytochrome.

Tetraspore In the Rhodophyta, the meiospore.

Tetrasporophyte In the Rhodophyta, the second diploid generation, produced by a carpospore and giving rise to tetraspores.

Textural Class Classification of a soil based upon the relative proportions of sand, silt, and clay present. There are 12 textural classes listed here in order from coarsest to finest: sand, loamy sand, sandy loam, loam, silt loam, silt, sandy clay loam, clay loam, silty clay loam, sandy clay, silty clay, and clay.

Texture The relative proportions of the various soil separates (i.e., sand, silt, clay) in a soil, as described by the classes of soil texture.

Thallus (thalli) Any plant body that is formed of parenchyma or plectenchyma but is not well-differentiated into organs.

Thatch Tightly intertwined layer of plant litter from accumulations of undecomposed or partially decomposed plant residues.

Theca General term for any stiff outer covering of a unicellular protist, and usually made up of interlocking plates. Dinoflagellates and diatoms are examples of protists with thecae.

Thermal cycler The machine used in PCR which can be programmed for the heating and cooling cycles.

Thermophilic eubacterium Type of bacterium found in a hot environment.

Thermotherapy Use of heat to reduce or eliminate pathogens in plant tissue; often used on plants prior to meristem culture to produce pathogen-free plants.

Threonine Nonessential amino acid found in cheese, eggs, milk, meat, legumes, nuts and seeds

Threshold velocity Threshold velocity is the wind speed at which particle movement is initiated.

Throughfall Rain that falls through and can be collected under the tree canopy of a forest. Throughfall interacts with leaves and materials present on leaves (e.g., dust, plant secretions, insect droppings, etc.) and therefore can be chemically very different from rain that falls directly to the surface.

Thylakoid disk Thylakoid disks are disk-shaped membrane structures in chloroplasts that contain chlorophyll. Chloroplasts are made up of stacks of thylakoid disks; a stack of thylakoid disks is called a granum. Photosynthesis (the production of ATP molecules from sunlight) takes place on thylakoid disks.

Thylakoid membranes A series of sealed membrane compartments found inside the chloroplast of plant cells. The membranes contain the photosynthetic apparatus and all of the chlorophyll in the leaf. Their structure as closed 'sacks' permits functional 'inside' and 'outside' spaces within the chloroplast across which phohosphorylation can generate the proton gradient necessary for ATP synthesis.

Thymine A pyrimidine found in DNA that base-pairs with adenine.

Thyroid hormone A hormone synthesized from tyrosine in the thyroid gland.

Thyrse A type of inflorescence having 3 pairs of cymes in 3 verticillasters

Ti plasmid A circle of DNA found in Agrobacterium tumifaciens that encodes genes for moving the plasmid into a plant cell and inserting a portion of the plasmid into the plant chromosome.

Tight junction A continuous network of protein strands around the circumference of epithelial cells, sealing the space between cells and forming a barrier between the apical and basolateral domains.

Till (glacial) A deposit of sediment formed under a glacier and left in place after the glacier's departure, consisting of an unlayered mixture of, silt, sand, and gravel ranging widely in size and shape.

Tillage The process of turning or stirring the soil.

Tiller A lateral shoot that develops from the axillary bud of leaves at the base of a cereal stem.

Tilth The physical condition of soil as related to its ease of tillage, fitness as a seedbed, and its impedance to seedling emergence and root penetration.

Time One of the conventionally accepted soil forming factors.

Tissue Group of cells, usually of similar structure, that perform the same or related functions.

Tissue culture Plant cells are grown in culture which allows them to be manipulated and then induced to develop into whole plants.

Titer Concentration of a virus.

Tolerance Ability of a plant to endure an infectious or noninfectious disease, adverse conditions, or chemical injury without serious damage or yield loss; (of pesticides) the amount of chemical reside legally permitted on an agricultural product entering commercial channels and usually measured in parts per million (ppm).

Tolerant The inherent ability of a plant to survive and reproduce after herbicide treatment. This implies that there was no selection or genetic manipulation to make the plant tolerant; it is naturally tolerant.

Tonoplast The vacuolar membrane in plant cells.

Topoisomerase An enzyme that catalyzes the reversible breakage and rejoining of DNA strands.

Topt The optimum temperature for an enzymatic reaction.

Totipotency The concept that even specialized cells contain all of the genetic information for an organism and, therefore, any cell should be able to regenerate into any tissue or into an entire plant.

Totipotent The ability of a single plant cell to grow, divide, and differentiate into an entire plant. Mammalian cells do not have this ability.

Toxicity Capacity of a substance to interfere with the vital processes of an organism.

Toxin In general, a poisonous substance of biological origin. Specifically, a compound produced by a microorganism which is toxic to other organisms.

Trace elements Nutrients which are required in very small amounts (plants or animals)

Tracer An identifiable substance, such as a dye, a radioactive isotope, or an inert chemical (i.e. bromide), that is introduced into a biological system and can be followed through the course of a process, providing information on the pattern of events in the process or on the redistribution of the parts or elements involved; this can also be referred to as a "label." For example, in a current study where calcium has been applied to an entire watershed, a tracer is being used to determine exactly where the calcium goes - how much is washed out by the stream, how much trees take up and where in these trees the calcium goes (leaves, bark, etc), and how much herbaceous plants, such as ferns, take up.

Tracheids One of the two major types of water-conducting cells found in plants (vessel elements are the others). The elongate cells are dead at maturity, have tapered ends and are arranged end to end and have walls that are partially coated with lignin.

Tracking dye A dye molecule visible in room light that is mixed with a sample. This allows the scientist to track the loading of the sample and indicate it's movement through an electrophoresis gel.

Trade Name The Trade Name is the name under which a product is marketed.

Trait The characteristic that results from an expressing gene(s). Ex. Upright leaves, drought tolerance, Bt resistance. A trait can be influenced by the environment.

Transcription The production of a complementary strand of RNA from a segment of DNA.

Transcription factor A DNA-binding protein required for the activation and regulation of expression of a gene.

Transcription The process by which the nucleotide sequence of DNA is copied into a single-stranded molecule of RNA. The nucleotide sequence of the RNA created is complementary to the DNA sequence except all thymine molecules are replaced with uracil molecules.

Transcriptional attenuation A regulatory mechanism that controls elongation of an mRNA molecule past a certain site on the DNA template.

Transcytosis The sorting and transport of proteins to different domains of the plasma membrane following endocytosis.

Transduce The process whereby a receptor initiates one or more sequences of biochemical reactions that connect the stimulus to a cellular response.

Transducin A G protein that stimulates cGMP phosphodiesterase when it is activated by rhodopsin.

Transduction The transfer of genes from one organism to another by viruses, especially in bacteria.

Transfection The introduction of a foreign gene into eukaryotic cells.

Transfer cells Specialized cells with intricate cell walls and large membrane surface thought to function in the transport of materials.

Transfer RNA (tRNA) RNA molecules that function as adaptors between amino acids and mRNA during protein synthesis.

Transform fault A type of plate boundary where two plates slide past each other.

Transformation The transfer of genetic materials from one organism to another by humans (genetic engineering); a means of genetic variation in bacteria by absorption and incorporation of DNA from another bacterial cell.

Transforming growth factor β (TGF-β) A polypeptide growth factor that generally inhibits animal cell proliferation.

Transforming Process of introducing foreign DNA into an organism which is then passed on to the organism's progeny.

Transgene A gene that has been genetically altered. They are usually used to transform organisms. A foreign gene that has been inserted into the host's genomic DNA.

Transgenic An organism that has a new genetically engineered DNA sequence found in every one of its cells. Genetically engineered organisms are transgenic. These two terms are used interchangeably.

Transgenic Plant A genetically modified plant with a new gene (transgene) that may impart a new function. For example, transgenic Bt corn can produce its own insecticide.

Transgressive segregation Segregation of progeny phenotypes that are more extreme than either parent.

Transhumance Farmers have a fixed residence but send herds and flocks (with attendant) to pastures

Transit peptide Special sequences found at the end of a protein to direct its transportation to the needed cellular location.

Translation The process following transcription during which the nucleotide sequence of mRNA is read and 'translated' into a chain of amino acids (protein). The mRNA sequence is read three nucleotides (codon) at a time, and each codon codes for a specific amino acid.

Translocation Transport of herbicide short or long distances. Herbicides movement from site of application to site of action might involve cell to cell transport or long distance transport in the xylem or phloem.

Translucent So clear that light may pass through.

Transmembrane path The route followed by water that sequentially enters a cell on one side, exits the cell on the other side, enters the next in the series, and so on. In this path, water crosses at least two membranes for each cell in its path.

Transmembrane proteins Integral membrane proteins that span the lipid bilayer and have portions exposed on both sides of the membrane.

Transmission electron microscopy See electron microscopy.

Transmit (transmission) To spread or transfer, as in spreading an infectious pathogen from plant to plant or from one plant generation to another.

Transpiration Process in which water is absorbed by the root systems of a plant, moves up through the plant, passes through pores (stomata) in leaves, and then evaporates into the atmosphere as water vapor.

Transponder An acoustic device that scientists place above the seafloor, usually in groups of two or more, to help them navigate deep submergence vehicles. To locate a transponder, a ping is broadcast from a ship to the seafloor. If the transponder hears the ping, it sends a reply ping, letting researchers know theyíre in the right place. The transponder and its batteries are protected inside a plastic case the size of a beach ball, and can operate from the seafloor for up to five years.

Transposition The movement of DNA sequences throughout the genome.

Transposon A DNA sequence that can move to different positions in the genome.

Trap crop Crop planted around a field to protect the inner crop from diseases transmitted by aerial vectors; host crop of a parasitic plant, such as witchweed (Striga spp.), that is planted to stimulate seed germination, and later sacrified by plowing under before the parasitic plant produces new seeds.

Traqueidas Uno de los dos tipos principales de células conductoras de agua en las plantas (los elementos de los vasos son el otro tipo); estas células alargadas mueren en la madurez, tienen extremos adelgazados donde se unen unas a otras y presentan paredes parcialmente cubiertas con lignina

Treatment In an experiment, the group to which an experimental manipulation has been applied; paired with a control group.

Tree core A small-diameter cylindrical sample removed from a tree trunk - from its surface all the way to its center - that does no permanent damage to the tree. It is possible to count the number of tree rings in cores, providing accurate estimates of tree age and growth patterns.

Trenching Physical separation of soil in a vertical plane to sever grafted roots between trees.

Triacylglycerol Three fatty acids linked to a glycerol molecule.

Triazines A family of herbicides with similar chemical structure and belonging to the photosynthetic inhibitors mode of action.

Trichocyst Organelle in ciliates and dinoflagellates which releases long filamentous proteins when the cell is disturbed. Used as a defense against would-be predators.

Trichogyne A female receptive hypha.

Trichome A plant epidermal hair, of which several types exist.

Triglyceride Sn-1,2,3-triacylglycerol

Trilete mark The three-branched ridge formed by the flattened facets of a meiospore from a tetrahedral tetrad.

Triple fusion nucleus In the Anthophyta, the triploid nucleus that results from fusion of one sperm nucleus with both polar nuclei.

Triplet codon A set of three nucleotide bases in DNA or RNA that code for an amino acid.

Triplet state A normally forbidden excited electron orbital that is reserved for an electron with the same spin as the electron remaining in the ground state. Upon excitation to an excited singlet state, there is some possibility that the electron will reverse spin. When this happens, the electron then drops down to the triplet state, which is at a lower energy level than the excited singlet state. Once in the triplet state, the electron must remain there until its spin is again reversed. Molecular oxygen is unusual in that it is normally in a triplet state.

Triploid Describing a nucleus or cell that has three times (3n) the haploid number (n) of chromosomes

Triploid fusion nucleus The result of the fusion of one male gamete (pollen) with two haploid female nuclei (polar cells) in the 'second polar event'. This nucleus divides to form the cells of the endosperm

tRNA (abbr. for transfer RNA) The RNA that moves amino acids to the ribosome to be placed in the order prescribed by the messenger RNA.

Trophic level Position in a food chain, determined by the number of energy-transfer steps to that level. A black-throated blue warbler is at a higher trophic level than the insect it has just eaten.

Tropomyosin A fibrous protein that binds actin filaments and regulates contraction by blocking the interaction of actin and myosin.

TS Transverse Section. The tissue sections were cut across the grain. The images presented are of sections from the middle of the grain (median).

Tube cells Cells formed from the inner epidermis of the pericarp in which the cells elongate parallel to the grain's long axis.

Tube nucleus In seed plant pollen, the nucleus that directs growth of the pollen tube.

Tuber An underground stem adapted for storage, typically produced at the end of a stolon.

Tubulin A cytoskeletal protein that polymerizes to form microtubules.

Tumor (gall) Abnormal swelling or localized outgrowth, often roughly spherical, produced by a plant as a result of attack by a fungus, bacterium, nematode, insect or other organism.

Tumor necrosis factor (TNF) A polypeptide growth factor that induces programmed cell death.

Tumor promoter A compound that leads to tumor development by stimulating cell proliferation.

Tumor suppressor gene A gene whose inactivation leads to tumor development.

Tumor virus A virus capable of causing cancer in animals or humans.

Turbidity The amount of solid particles that are suspended in water and that cause light rays shining through the water to scatter. Thus, turbidity makes the water cloudy or even opaque in extreme cases.

Turgidity State of being rigid or swollen as a result of internal water pressure.

Turgor pressure The internal hydrostatic pressure within plant cells.

Turnover number The number of reactions per second that the enzyme can catalyze under optimum conditions.

Tylose, Tylosis A balloon-like outgrowth from a parenchyma cell that expands through a pit in a xylem vessel wall and into the lumen of the vessel, either blocking it completely or partially.

Type The example on which the description of a scientific name is based, and which fixes the application of the name (the type genus of a family, or the type species of a genus).

Tyrosine Nonessential amino acid

U

Ubiquinone See coenzyme Q.

Ubiquitin A highly conserved protein that acts as a marker to target other cellular proteins for rapid degradation.

Ultrastructure The detailed structure of a specimen, such as a **cell**, **tissue**, or **organ**, that can be observed only by electron microscopy. Also called fine structure.

Umbel An indeterminate inflorescence, usually flat-ped with peidcels arising from a single point, like an umbrella.

Armitage, A.; Herbaceous Perennial Plants

Unaltered fossils fossils which retain more or less their original chemical and structural composition; most commonly shells of calcite (mollusks) or silica (diatoms).

Undifferentiated Cells that have not developed into specialized tissues. Undifferentiated cells, called callus, are used in tissue culture during the transformation process.

Undulipodium Another term for a eukaryotic flagellum.

Unicellular Of organisms, consisting of one cell per organism.

Uniflagellate Having one flagellum.

Uninucleate Having one nucleus .

Uniport The transport of a single molecule across a membrane.

Unitunicate Having a single ascus wall.

Universal code The theory that the genetic code is used by all forms of life. This is what allows DNA from different organisms to be introduced into a new organism and function correctly.

Uptake The process of getting a herbicide from the outer surface of a plant to the inside of the plant where the herbicide will either induce its herbicidal activity or be translocated.

Uracil A pyrimidine found in RNA that base-pairs with adenine.

Uranium series dating A method of dating material that contains elements that have been formed by radioactive decay of uranium.

Urea Dry nitrogen material produced by reacting ammonia with carbon dioxide; contains the highest percentage of nitrogen of the commonly used dry fertilizers and is rapidly replacing ammonium nitrate. When surface applied, urea is the most rapidly volatilized of the dry nitrogen materials.

Urea-ammonium nitrate (UAN) A non-pressure solution of ammonium nitrate, urea, and water. Two grades are most common: 28-0-0 (10.7 lbs/gal) and 32-0-0 (11.1 lbs/gal).

Urediniospore (also urediospore, uredospore) The asexual, dikaryotic, often rusty-colored spore of a rust fungus, produced in a structure called a uredinium; the "repeating stage" of a heteroecious rust fungus, i.e. capable of infecting the host plant on which it is produced.

Uredinium Fruiting body (sorus) of rust fungi that produces urediniospores.

Utilization The proportion of current year's forage production that is consumed or destroyed by grazing animals.

V

Vacuolate cell A cell derived from a progenitor cell during embryogenesis and having a large vacuole. Vacuolate .cells usually are larger than cylasmic cells which result from the same cell division.

Vacuole A large membrane-enclosed sac in the cytoplasm of eukaryotic cells. In plant cells, vacuoles function to store nutrients and waste products, to degrade macromolecules, and to maintain turgor pressure.

Vadose zone Zone of material between the surface and the water table.

Vapor A substance in the gaseous state

Variability The quality of being likely to change or vary over time; lack of uniformity. There are two types of variability: temporal and spatial.

Variable A quantity able to assume different numerical values

Variegation Pattern of two or more colors in a plant part, as in a green and white leaf.

Variety Crop plants within a species that have the same genetic composition. Because plants in a variety are usually heterozygous, their offspring will not remain genetically pure (i.e., corn hybrid varieties).

Vascular Pertaining to any plant tissue or region consisting of fluid-conducting tissue; for example, xylem and phloem. The term is also sometimes applied to a pathogen that grows primarily in the conducting tissues of the plant.

Vascular bundle Strand of conductive tissue, usually composed of xylem and phloem (in leaves, small bundles are called veins).

Vascular cambium A layer of meristematic cells located between xylem and phloem tissues which divide to give off cells toward both tissues. This process forms secondary growth in stems and roots in broadleaf plants.

Vascular cylinder (stele) The cylinder of vascular tissue in stems or roots.

Vascular plant A plant that produces vascular tissue.

Vascular tissue A tissue system which transports water and nutrients (called xylem) and photosynthates (called phloem) around the plant.

Vascular tissue Collectively, xylem and phloem.

Vascular Wilt A disease in which the pathogen is confined to the vascular system of the host and in which wilting is a characteristic symptom; plants lose their turgidity and become flaccid, leaves collapse.

Vector A living organism (e.g., insect, mite, bird, higher animal, nematode, parasitic plant, human) able to carry and transmit a pathogen and disseminate disease;

Vegetative Referring to somatic or asexual parts of a plant, which are not involved in sexual reproduction.

Vegetative propagation Asexual reproduction; in plants, the use of cuttings, bulbs, tubers, and other vegetative plant parts to grow new plants.

Vein Small vascular bundle in a leaf.

Vein banding Symptom of virus disease in which regions along veins are darker green than tissue between veins.

Vein clearing Disappearance of green color in or around leaf veins.

Ventral groove Also called the crease The furrow on the ventral side of the grain. The tissues of the ventral groove are adapted for efficient transfer of nutrients.

Vertical resistance (specific resistance) Resistance which is effective against some biotypes or races of the pathogen, but not others, usually inherited monogenically and expressed qualitatively (see general resistance, horizontal resistance, race non-specific resistance).

Verticillaster A false whorl of flowers

Verticillate Arranged in whorls. Armitage, A;

Herbaceous Perennial Plants

Verticils A whorl of flowers or leaves

Vertifolia effect A term used to describe the loss of general (horizontal, minor gene, multigenic, polygenic) resistance in a cultivar after several generations of selection during which a major gene confers resistance to the dominant race or biotype of the pathogen; first observed in the potato cultivar Vertifolia with late blight resistance.

Vesicle Thin sac in which zoospores are differentiated and released;

Vessels Or vessel elements, one of the two major types of water-conducting cells found in plants (tracheids are the others). The elongate cells are dead at maturity, have tapered ends and are arranged end to end and have walls that are partially coated with lignin

Viability The measure of the number of individuals in a given phenotypic class that survive relative to another class.

Vinblastine A drug that inhibits microtubule polymerization.

Vincristine A drug that inhibits microtubule polymerization.

Virescence State or condition in which normally white or colored tissues (e.g. flower petals) become green.

Virion Complete virus particle.

Viroid An infectious, nonencapsidated (naked) circular, single-stranded RNA.

Viroplasm Cellular inclusions that are sites of synthesis of viral components and the assembly of virus particles.

Virulence The degree or measure of pathogenicity of a given pathogen; relative capacity to cause disease.

Virulent Highly pathogenic; having the capacity to cause severe disease .

Viruliferous Virus-laden, usually applied to insects or nematodes as vectors.

Virus A submicroscopic, non-cellular structure consisting of a core of infectious nucleic acid (either RNA or DNA) within a protein coat.

Viscin Sticky substance produced on the seed coat of parasitic flowering plants in the Viscaceae (mistletoes) that helps attach the seed to the host plant branch.

Vm The limiting velocity of an enzymatic reaction. This would occur at an infinite substrate concentration.

Volatile Liquids or solids with relatively high vapor pressure at normal temperatures. The higher the vapor pressure, the more likely the substance will vaporize into a gas.

Volatility The ability of a compound to change from a liquid phase to its gaseous phase, dependant on chemical properties of compound and environmental conditions.

Volatilization The conversion of ammonium into ammonia gas.

NH4 – – – – > NH3(gas)

Volcanic eruption Ejection of hot steam, molten rock, and/or rock fragments through openings in the earth's surface.

Volunteer Self-set plant; plant seeded by chance.

W

Walling-off Separation of diseased from healthy tissues by barrier tissues produced by a diseased plant.

Water holding capacity The maximum amount of water a soil can hold against the pull of gravity.

Water filled porosity The percent of water occupying the empty spaces of soil and is inversely related to the percent of air occupying the empty spaces of soil. Total porosity= air-filled porosity plus water-filled porosity.

Water mass A body of seawater that shares the same origin. Each water mass has characteristic temperatures and salinities.

Water potential The chemical potential of water; a measure of the ability of a substance to absorb or release water relative to another substance

Water ripe (also called Pre-milk) A description the newly pollinated grain. When punctured the grain contents are very watery.

Water sprout Small, rapidly growing shoot or branch on a large stem, developed from adventitious tissues.

Water table The upper surface of ground water or that level in the ground where the water is at atmospheric pressure.

Water yield Amount of water leaving a watershed, as measured at the weir. Given in units of height per area (e.g, mm water per square meter).

Water holding capacity The amount of water held in soil pores against the pull of gravity.

Watershed The drainage area of a stream, river, or other body of water.

Wavelength The distance between equivalent positions in the electromagnetic wave formed by a photon as it moves through space. For visible photons, the wavelengths are in the range of 400 to 700 nm.

Weaning Removal of young mammals from their source of milk

Weathering The complex combination of physical, chemical and biological processes that decompose, disintegrate and alter rocks and minerals at or near the Earth's surface. Weathering can be subdivided into chemical weathering, physical weathering, and biological weathering. (Soil and Environmental Science Dictionary)

Western blot A technique in which proteins are separated using electrophoresis and transferred to a membrane so they can be identified and studied.

Wetland A transitional area between aquatic and terrestrial ecosystems that is inundated or saturated for long enough periods to produce hydric soils and support hydrophytic vegetation.

White rot (of wood) Type of wood decay resulting from enzymatic action of fungi; it degrades all components of wood, including lignin, leaving the wood light-colored and spongy.

White rust Common name usually used for diseases caused by oomycetes in the genus Albugo.

Whole tree harvesting A form of logging in which entire trees, including branches and crowns, are removed from a site. An experimental whole tree harvest was conducted on Watershed 5.

Whorl Circular arrangement of like parts.

Wild type The phenotype characteristic of the majority of individuals of a species under natural conditions.

Wilt Drooping of leaves and stems from lack of water (inadequate water supply or excessive transpiration);

Wilting point The moisture content of a soil at which most plants wilt.The wilting point is commonly estimated by measuring the 15-bar or 1.5 MPa percentage of a soil.

Wind break A row or other grouping of trees or shrubs used to provide protection against the effects of high velocity winds.

Winterburn Foliar necrosis, often marginal, of plants that retain their leaves in winter due to water deficiency because they cannot take up water from frozen soils.

Witches' broom Disease symptom characterized by an abnormal, massed, brushlike development of many weak shoots arising at or close to the same point.

Wollastonite (CaSiO3) A mineral containing calcium silicate. Calcium was applied to the watershed in the form of pelletized Wollastonite.

Wood Secondary xylem.

Woody caudex The swollen stem base of certain nonwoody perennial plants that survives over the winter and from which new growth is produced.

Wound Injuries to plant tissues that often breach barriers (cuticle, bark, cell walls) that might otherwise exclude pathogens; some pathogens (e.g. viruses) can enter plants only through a wound; wounds may occur from natural growth processes, physical and chemical agents, animals (especially insects), and many human agricultural activities, such as pruning.

X

Xanthophyll A class of fat-soluble photosynthetic accessory pigments.

Xenobiotic Foreign chemicals; natural or synthetic substances that cannot be utilized by plants for energy-yielding processes.

Xeric Dry; applied to features of plants or places where they grow.

XLB (xylem-limited fastidious bacteria) Bacterial pathogens of plants found only in the xylem, causing wilt, scorch, and stunting symptoms; vectored by insects that feed on xylem fluid; not sap transmissible; require complex culture media.

X-ray crystallography A method in which the diffraction pattern of X rays is used to determine the arrangement of individual atoms within a molecule.

Xylem A complex vascular tissue through which water and minerals are conducted through the plant. Xylem also aids in the structural support of a plant as the primary component of its cell walls is lignin.

Xylem-limited fastidious bacteria (XLB) Bacterial pathogens of plants found only in the xylem, causing wilt, scorch, and stunting symptoms; vectored by insects that feed on xylem fluid; not sap transmissible; require complex culture media.

Y

Yeast Unicellular ascomycetous fungus that reproduces asexually by budding .

Yeast artificial chromosome (YAC) A vector that can replicate as a chromosome in yeast cells and can accommodate very large DNA inserts (hundreds of kb).

Yellows Disease characterized by chlorosis and stunting of the host plant .

Yield An aggregate of the products resulting from growth or cultivation .

Yield drag A negative effect on grain yield associated with crop plants that have a specific gene or a specific trait.

Yield lag A relative reduction in yield observed in some hybrids or varieties compared with the yield observed in the most recently produced hybrids or varieties.

Yield potential The highest yield a plant (hybrid, variety, etc.) is capable of producing when grown in ideal conditions.

Z

Zero grazing Where forage is cut and transported to livestock

Zinc carbonate (52-56 percent zinc) Inorganic compounds; very insoluble in water; not effective zinc sources if granulated, but effective if very finely ground.

Zinc chelate (Zn-EDTA most common) Blended with dry fertilizers, applied in a row band, granular zinc chelate is likely to be more effective than granular zinc sulfate because of its mobility.

Zinc finger domain A type of DNA binding domain consisting of loops containing cysteine and histidine residues that bind zinc ions.

Zinc oxide (50-80 percent zinc) Inorganic compounds; very insoluble in water; not effective sources of zinc if granulated, but effective when finely ground.

Zinc sulfate Most commonly used dry zinc material. Relatively water-soluble inorganic compound.

Zonate Targetlike development of tree canker, characterized by successive, perennial rings of callus; any symptom appearing in concentric rings .

Zoonoses Diseases transferrable between humans and animals.

Zooplankton Microscopic animal organisms that drift in bodies of water.

Zoosporangium Sporangium, or spore case, that bears zoospores.

Zoospore Fungal spore with flagella, capable of locomotion in water.

Zwitterionic Having both a positive and negative charge.

Zygomycetes A fungus group, also called the Zygomycota, characterized by nonseptate hyphae, sexual zygospores, and sporangiospores produced in a sporangium; common post-harvest pathogens.

Zygospore Sexual resting spore formed from the union of gametangia in the Zygomycetes.

Zygote The cell formed when the egg and sperm fuse. The cell has two copies of each chromosome (2n) and will continually divide to develop into an entire organism.